TIME AND SPACE
THE CONCEPTUAL ANSWER

COPYRIGHT: CLINICAL PRESS LTD 1995

All rights reserved. No part of this publication may be reproduced, stored in a retrieval system or transmitted in any form or by any means electronic, mechanical, photocopying or otherwise, without the prior permission of the copyright owner.

Published by

Clinical Press Limited
Registered Office, Redland Green Farm
Redland Green, Redland, Bristol, BS6 7HF

Butler P.,   Irvine A.,   Coral A,   Blease S.

Time and Space, a conceptual answer

Goddard, Paul Richard

ISBN 185457 035 8

TIME AND SPACE
THE CONCEPTUAL ANSWER

by

PAUL RICHARD GODDARD BSC., MBBS, MD, DMRD, FRCR

CLINICAL PRESS

Acknowledgements

The people who have helped me with this theory are numerous. Some would be pleased to be mentioned but others, perhaps, would not.
The most important have been Jeremy Goddard, Nigel Burnell and
Dr. Alan MacKenzie. But always supportive and willing to return the discussion in a lively manner were Lois and Mark Goddard.

Professor Howell Peregrine, Professor Graeme Bydder and Dr Peter Dawson have provided specific assistance and Jeremy Mann a sympathetic ear. Iain Watt, Simon Freeman, Charles Wakeley, Simon Blease, Jim Virjee, Mike Halliwell, Mike Keen, Professor Peter Wells and everybody around the BRI have joined in with the discussion in a spirited manner.

A special thank you to Steve Atkinson for pointing out the necessity for balance between order and chaos.

They do not all accept the validity of the theory. But healthy scepticism is what is required and a desire to test the theory.

Time and Space
The Conceptual Answer to the Grand unifying Theorem

Preface

Part 1

Chapter 1: The ancients
Chapter 2 : The age of reason
Chapter 3: Space
Chapter 4: Time
Chapter 5: Bar magnets
Chapter 6: Quantum theory
Chapter 7 Special Theory of Relativity,
Chapter 8 General Relativity
Interlude: The Horse's Mouth

Part 2 Goddard's Conceptual Grand Unifying Theorem

Chapter 9 Introduction to Part Two
Chapter 10 Elementary Rules of Geometry for 4,5 and 6 Spatial Dimensions and a Law of Duality, Complementarity and Distance Probability
Chapter 11 The important whimper, the parasitic universe and the self-replicating particle
Chapter 12 Fridge magnets and superstring: unlocking the secrets of the Universe
Chpater 13 What is Gravity ?
Chapter 14 Time and motion studies
Interlude: Energy?
Chapter 15 Quantum mechanics
Chapter 16 Weak interaction, strong interaction and the Higg's boson
Chapter 17 Mathematics, the Mandelbrot Set and the Pattern of the Universe
Chapter 18 The Fractal Universe from Galaxy to Quark
Summing up and where do we go from here ?
Glossary

VI

TIME AND SPACE : THE CONCEPTUAL ANSWER

A THESIS ON PHYSICS AND COSMOLOGY PROVIDING THE NECESSARY BUILDING BLOCKS FOR A GRAND UNIFIED THEOREM

Preface

People , by and large, don't read the preface until they have finished reading a book, enjoyed it and decided they want to read a bit more. So, if that is the case for you : hi there, I double guessed you! There is nothing of real importance in this preface that you need to read before you start. However I would like to make a few points. Many physicists, mathematicians, cosmologists etc. will be disappointed with this book after all the hype that has sold it. The really disappointed ones will not read this preface because they will already be fed up with what I am saying.

The reason they will not like it is because it is written in **English** not in mathematics. The physicists feel that the concepts behind the universe must be astonishingly complex because the universe itself is so complex. But consider DNA: the concept of DNA, a self-replicating spiral molecule is not difficult but the human being that arises is possibly the most complex structure in the galaxy and maybe in the entire universe.

Others will have an open mind and I would like to assure the reader that there is nothing in this book that is illogical. In some places I have used mathematics but only where I feel I have to try extra hard to get the physicists on my side : on each such occasion I have provided the same information in standard understandable English. I have argued from syllogistic logic and perhaps now is the point to express what this means :
A typical such argument is as follows:

All men are mortal
I am a man
Therefore I am mortal

This is a correct form of logic

You **cannot** reverse a syllogism. You are not permitted to say:
All men are mortal
I am mortal
Therefore I am a man.
This is illogical (I might be a cat or a fish etc. etc.though whether I could argue with any form of logic is another question)

You **can** say
All men are mortal
If I am a man I must be mortal

and you **can** say
If all men are mortal
and **if** I am a man

Then I must be mortal
and
All men are mortal
If I am a man, then I must be mortal.

Since the weakness of syllogistic argument lies in its original two statemtents (or premises as they are known) the latter form of argument is probably the only acceptable one when dealing with problems of cosmological study.

Since we cannot be certain we are right , Science proceeds by an idea or postulate that appears to be correct being put forward and tested in two ways:
Firstly ... does it explain or include the knowledge that we already have
Secondly ... can it be tested by any observable experiment.

The ideas put forward in this book, although apparently contradicting much that we have learned by rote, could certainly be used to explain current knowledge in Cosmology and Physics and various tests have been proposed. The ideas are, in fact not mine .... I thought that they were but I keep finding references to them in different places. All I have done is draw them together. Whether they explain philosophy and religion is a different question since this is not generally open to testing and the verdict may be "we cannot know"

Certainly the conclusions drawn are astounding but then astounding ideas are required to explain and draw together the advances that have been made in Science by dedicated workers over the last 90 years.

The background to this work

The conceptual answer to the Grand unifying theorem ?

A difficult question

"Why has a doctor of medicine written a book on physics ?" This may well be the question on your mind when you pick up this book, assuming of course that you do. Well he hasn't yet, I am only just beginning. .... and like all good stories I have decided to start at the beginning.

I suppose it started back at school when I was aged around fourteen or fifteen. I had heard about Leonardo da Vinci and that he was the man who encompassed all human knowledge. The things he invented, his art and his science and the advice he gave to governments and to individuals was amazing and there and then I decided to encompass all human knowledge. This turned out to be fairly straight forward at the O-level stage (our GCSEs), at least in mathematics, physics, chemistry, biology and geography. It turned out to be impossible in French (I am not a linguist) and I therefore gave myself a let-out ... I would not worry about languages since I thought that learning a different language would only permit me to express the same thoughts in other tongues. There is also the story of the Tower of Babel to consider. In this idea I am hopelessly wrong and there is no doubt that languages permit the expression of different ideas, poetry and literature and are an invaluable well of human experience, but I am afraid they are not for me (unless I go to live in a different country, of course!).

I had decided to go into Medicine since I thought that this was the best way of using science, at which I had a talent, and helping people directly.
Onto A levels and I grasped all the concepts of Physics and Chemistry immediately .... but Zoology ...oh dear ! This seemed to be rich in description and poor in concept and therefore was a memory game and could not be understood in a few logical thoughts. I passed the Physics and Chemistry with flying colours but had a low pass in Zoology. Off to medical school. This was even worse than zoology ! The memory game in subjects like anatomy was appalling and there was no way that anybody could ever understand it all. I had failed my task at the first hurdle.

I had consciously stopped trying to emulate Leonardo but it seems that my subconscious had not given up or perhaps its just the way I am. Whilst I was at medical school (supported by my parents love and money !) I also became very interested in music and having played the piano since the age of six, learnt how to play blues and jazz. Again it was a question of looking for the hidden concepts behind the apparently random but obviously pleasing sounds and I was delighted to discover that there were some simple rules and I was able to teach them to my brother (an excellent musician) and friends. I will never play jazz like the truly gifted performer but I have managed to lead a succesful jazz band (Dr Jazz) for the last ten years and we have recorded a single, a CD and a tape and sold them all for charity.

Medical school seemed to me to be a struggle of stupendous proportion in which I had to try and cram in as many "facts" as possible. The realisation that many of these facts were contradictory did nothing to help me. But my girlfriend, Lois, certainly did. Her recollection of that time is that I should have been applying myself more to medical study, but there you go.

After qualifying and whilst doing my housejobs I saw an article in Scientific American about computerised axial tomography scanning ... CAT scanning as it was called in those days .... and I decided that radiology was a potential career. Lois, now my wife, qualified as a dentist in 1975 (with distinction and honours in her BDS .... something I failed to achieve) and in 1977, I started in Radiology in Bristol. This was a great time ... a reawakening of my interest in physics, learning new subjects such as radiography and radiological techniques and for the first time learning the anatomy properly. In addition I started to do an MD thesis under the tutelage of the real intellect in our department, Iain Watt.. This also led me into the exploration of physics and new concepts such as a pixel (picture cell) and a voxel (volume cell ..... no, not the old Viva... that's a Vauxhall !).
All of this is important when one considers the solutions to the problems besetting cosmologists, physicists, etc. but we will come to this anon...... this book is like a thriller .... all the pieces must be in place and the answers will be delivered as we go along but the final denouement does not occur till fairly near the end and is followed by the summing up for Watson..

In 1982 I was offered a job in Manchester running a new machine for research purposes. The job was offered to me by one of the greatest living radiologists, Professor Ian Isherwood, and the machine was a Magnetic Resonance Imaging scanner (MRI scanner). I turned the job down because I was shortlisted for a post in Bristol and I had been assured that we would be getting a machine here. Perhaps I was wrong to turn down the Manchester job and perhaps this book would have been written earlier if I had gone there but we will never know.

The funding for the MRI machine in Bristol never transpired so I set about raising the money myself. I set up a committee in 1984 and this led to the establishment of the Bristol MRI Scanner Fund in 1985. Our patron and first large donor was John James who gave the machine as an outright gift costing £900,000. A further £2.6 million has been raised through the generosity of the people of Bristol and the South West. My one regret was that the fund committee never saw it as their task to fund basic research ... I was part of that committee but could not persuade them. But the clinical work and clinical research we did was extraordinary ... a golden period of opportunity followed. The trust committee may have been right since our approach to clinical MRI certainly altered the thinking on MRI throughout the country.

Once again I came across new concepts and these were somewhat more complex. For the first time quantum phenomena were noticeably affecting the picture at all levels .... well, I lie, not quite the first time ..... quantum mottling is an obvious feature on CT (the name of CAT scanning changed to computed tomography, a better choice, under the influence of the Professor in Radiology in Bristol, Professor Rhys Davies) and on fast plain X-ray films if one turns down the wick too much (i.e. a low intensity of beam).

I started to learn simple concepts of MRI and teach them to my friends and colleagues. They were simple concepts quite often and, as is the way when you find a simple concept, once having told the confused person the correct approach they reply ... "I knew that all along !" ..... and indeed they did, somewhere in their brain they had included the idea as a possibility but it may not have been able to surface. Or perhaps the really basic concepts are so strikingly correct that some people grasp them so quickly that they believe they knew it all along or they would not otherwise have understood what you were saying . More of this in the section on creative thinking.

In 1988 I published a special issue of the Bristol Medico-Chirurgical Journal on MRI and included in this an account of the physics of MRI and one of the truly great figures in the subject, Dr. Graeme Bydder (now Professor at the Hammersmith Hospital) was led to remark to me that I had blown the gaffe on the subject of MRI by making it too simple !

But it still did not seem simple to me and still doesn't ! This , of course, led on to a study of quantum mechanics.

But I have left out one salient and very important fact. In 1980 my first son Jeremy

was born, followed in 1984 by Mark. Both are a source of great delight and refreshing ideas (how does a corkscrew work, Dad ? asks Mark .... and he wants a sensible answer). In 1984 I was sitting on the settee with Jeremy watching television .... not a great source of new ideas normally in my opinion since it uses too many of the senses and leaves too little to the imagination ..... but on this one occasion something stirred Jeremy to ask me the important question ...
"Dad" "Yes, Jeremy" " What if I took a spaceship out into space and went past the moon" .....I thought he was going to say would I plunge into the sun but he didn't say that "and I went past all the planets and then out past all the stars **what would there be beyond all the stars** ?" At the age of four he is asking me a question to which I have no sensible answer .... with all my study I can't even answer the first difficult question !

So read on for the answer that was eleven years in coming ....

CHAPTER 1

The ancients

Introduction

At one time there was no difference between the artist and the scientist. The philosopher was the logician and the mathematician, the poet was the physicist. The person who had new ideas was listened to with great respect by his followers and with considerable alarm from the established hierarchy. The ancient philosophers were also poets and artists and they tried to find the overall balance in nature. With the power of their minds they managed to provide amazing insights into the world.

Superficially this has all changed. We live in a scientific age with new technology and a precise knowledge of measurement. In fact the latter has become so important that the statement has been made by a number of eminent scientists that " There is no science without measurement". In our present limited idea of science, that may be true. But have things really changed ? Do we really believe that in just a few hundred years people's minds have changed from being disordered and jumbled into a new way of rational thinking that was not there before and is gloriously epitomised in the modern scientist with all the accompanying wonderful achievements ? After all the scientists have achieved amazing order in all things : we can even measure the size of the universe and the size of the smallest particles. If this is the case why is it that scientists are disliked by so many people ?

If you probe more deeply you will find that the large divide between science and art is not the case at the individual level. The person who excels in science is often also excellent at music and poetry but he may well not let people know this in order to avoid ridicule. A great friend of mine who I admire for his radiological ability told me a few years ago "Paul, you will never really be number one at anything because you try to be good at all things. You will be Jack of all trades but master of none". This may be true in my case but it is far from being a universal truth. The great scientist could be a great thinker in a large number of disciplines and he often is.

Unfortunately the reverse is too often the case. Too many people with a "scientific bent" give up trying to understand the arts as well. Moreover so many children with an artistic streak and immense talent have given up struggling to understand science at a very early age and have decided that there was no sense in it (for them) and they have then turned to the arts , closing their minds to science. This is an immense pity because such people, lawyers, poets, artists could see the beauty of science and the excitement if they tried. They would also be more able to spot the total illogicality that can occur and would have more open minds than most scientists. Luckily there are still some scientific artists and artistic scientists and I would cite architects and engineers as examples...... when they let their artistic talents complement their scientific talents the results are glorious but when they do not the results are high rise blocks of flats.

The desire for measurement at all costs has pervaded society leading to absurdities such as the directive from the European Community which set standards on how long and how curved bananas should be. You can think of many more examples of such absurdities.

Also pervading society is a feeling of uncertainty and that nothing is permanent, all things are purely relative. There is no surprise to find that these much overused words are the cornerstone of the three major theories that have dominated this century. These are, of course Quantum Mechanics, Special Relativity and General Relativity. There is also a widespread feeling that nothing really matters and some even feel that there is no true reality and that time is purely a psychological myth. This also emanated from scientific theory which, using equations, can show that reactions should always be able to go backwards as well as forwards and when they do not it is the exception . That everything changes in time and never reverses is usually ignored in traditional science.

What about Religion ? Could it really be that so many great thinkers over the centuries are completely wrong and that we, at the end of the second millenium, are the first human beings to get it right? Is it true that Religion has nothing to offer the modern world or is there incorporated into each Religion a kernel of truth that should be sought ? Could there be more to this life than pure material gain ..... a spiritual side that we cannot see ?

If all of our advances are so marvellous and our science is so clever why do we hanker for the ancient days when life was more simple ? Is it simply nostalgia?

These questions will be pursued in the later chapters, but I would put in a word of caution ...... Socrates was made to take poison and Galileo was put on house arrest. In the ancient times they were not all tolerant of new thinkers. Moreover there was no modern medicine and living conditions for the majority were squalid. The advances of modern science have been astounding and we should try to keep it all in perspective.

Socrates, Aristotle, Plato and others

The ancient Egyptians and Greeks had an amazing ability to discover facts about the world without the need for sophisticated equipment. The Mayans of South America had an obsession with measuring time and had a calendar that was far more accurate than our own.

In 235 BC Eratosthenes of the University of Alexandria was able to measure the circumference of the Earth. He knew the world was round and divided the globe into frigid, temperate and torrid zones. Other ancients stated that the earth moved round the sun and they were able to measure the diameter of the moon within 5% of the value that is now accepted.

Aristotle, Socrates and Plato were amongst many of the philosophers who did not restrict themselves to moral issues but made amazing advances in science.

Aristotle, who lived between 384 and 322 BC tried to systematize knowledge in the way that Euclid had systematized geometry and it is Aristotle's systematic approach that became the method from which science later arose. Unfortunately his followers considered his words to be the final word on the subject ..... something that he had never intended. This has been the way for all great scientists throughout history. Galileo, Newton, Planck and Einstein all thought that their science was the best way of understanding the information available but that a better understanding would be possible in the future. In every case the followers have obscured the principles by slavishly believing in the original master. The same can be said of many religions and of the arts...... it seems to be the normal human action to accept one idea and then to rigidly believe it. More can be said about this later but there is a ferment in thought at this moment permitting new interpretations of old information and theories.

The ancients studied the world in great depth with just thought and the power of logic (see the preface). They looked at matter and came up with two possible answers : either it was made up of indivisible particles (atoms), or it was infinitely divisible.

We know now that matter is made of atoms, but that they are divisible. How divisible are they, is the quark the ultimate "atom" ?

They had more trouble with motion. Aristotle divided motion into two main classes: natural motion and violent motion. In Aristotle's view to be at rest was the natural state and each object had a proper place. Aristotle thought that rock came from the ground and therefore wanted to go back there, feathers came from air and ground and slowly fell to the ground, smoke came from the air and therefore rose back into it. Violent motion occurred when there was something pushing or pulling the object. Aristotle believed that all motions resulted either from the nature of the object or from a sustained push or pull and that an object would stop moving if you stopped pushing it unless some other action kept it moving. Aristotle's views on motion were wrong but we would do well to consider the same questions again today.

*We know that an object will keep moving once it has acquired a velocity but **why** does it keep moving ?*

Chapter 2

The age of reason

Galileo and Newton

Galileo (1564-1642), as well as studying the stars, made two major contributions to science :
1) Falling objects fall to the ground with the same acceleration whatever their weight unless air resistance has a major influence. This explained the difference between a feather and a stone.
2) If you set a body in motion it will continue to move in a straight line until a force stops it ...... it has momentum

Newton further formulated the laws of gravity and motion

Newtons first law of motion
Every material object continues in its state of rest, or of uniform motion in a straight line, unless compelled to change that state by forces imposed upon it.
In other words an object keeps doing what it was doing unless a force acts upon it to change it.
If a body is not moving it is said to be at rest and has inertia.
If a body is moving it will keep moving unless a force opposes it. The propensity (or desire) to keep moving is known as momentum

We might bring in our first mention of dimensions at this point. The dimensions are the magnitude of a particular factor. The particular factors are basically Mass, Length and Time.

A mass that is stationary has inertia and the dimension of Mass (M).

A mass that is moving has a velocity (a speed in a single direction) and is said to posess momentum. The dimensions of momentum are its mass multiplied by its velocity. Velocity has the dimensions of Length over Time (as in miles per hour and metres per second to mix Imperial and SI units). So the dimensions of momentum are

Momentum = Mass x velocity
$$= M \cdot \frac{L}{T}$$

So when a mass has momentum it has gained dimensions of length and time.

(Remember that momentum is a concept but that the gaining of the dimensions of length per time was an observation)

Newton's Second Law of Motion

The acceleration of an object is directly proportional to the net force acting on the object, is in the direction of the net force and is inversely proportional to the mass of the object.

This simply means that the acceleration is in the direction of the force and can also be written as

Force = mass x acceleration. The dimensions of acceleration are length per second per second.

Force = Mass . $\dfrac{L}{T^2}$

So an accelerating mass moving from rest acquires dimensions of $\dfrac{L}{T^2}$

The Classical Newtonian concept is that a mass will not have a velocity unless it is accelerated by a force and if a force is being applied the body will continue to accelerate. Whilst. it is being accelerated it will have dimensions of "Energy"

E = mass x velocity^2 = $m \dfrac{L^2}{T^2}$

You will have gained a further dimension of length.
After you use a set time for the acceleration you will have used energy. Energy in the past tense is known as work done and this will have the same dimensions as energy.

You then stop applying the force the body will continue with a velocity and will have momentum which as we already know has dimensions $m . \dfrac{L}{T}$

So the body with velocity has lost a dimension of length (L) divided by a dimension of time. (T)

Questions. Where do the dimensions go to when you have used them ?

The dimensions of energy are $m \dfrac{L^2}{T^2}$

or mL^2T^{-2}

If m = mass and L^2 is area, what does T^{-2} mean. What is time squared ?

The physicist may reply that this is a meaningless question and you can multiply time by itself as often as you like. Why is this the case ?

(Remember that energy is the concept but the gaining of L^2T^{-2} was an observation)

Newtons Third Law states that : Whenever one object exerts a force on a second object, the second object exerts an equal and opposite force on the first . This is usually thought of as "for every action there is an equal and opposite reaction"

This is the hardest law to understand and you will see why in a minute

Now follows an imaginary discourse between Socrates and Newton

Socrates
You tell me that things move due to something called a force and that the Gravity of the Earth is one such force

Newton
Yes that is right and the Earth exerts gravity on a body. The gravity exerted depends on the mass of the object

Socrates
If this third law were true then the force of gravity would be balanced by an equal and opposite force (antigravity) and objects could not fall if you dropped them in the air. They would float due to the antigravity.
But objects do not float in mid-air .... they fall with a crash.

Newton
You misunderstand the law. The falling object exerts an equal force on the earth and the combined pull draws them together

Socrates
Thats not an equal and opposite reaction. That is double the action pulling them together

Newton
You've got me wrong again.
The equal and opposite reaction is due to air resistance .

Socrates
So air resistance will stop the object and make it float in the air if it falls for long enough.

Newton
Air resistance is dependent on velocity and the net force (gravity minus air resistance) will eventually be zero at which time the object will reach terminal velocity and will no longer accelerate. If the object were to slow down it would of course experience less air resistance and would speed up again to the terminal velocity. It cannot speed up more because there is no net force accelerating it.

Socrates
Why did this equal and opposite reaction take so long to come into action?

Newton
Let me try to explain it again. The object could be resting on the ground. The object is pulled by the Earth and therefore exerts an equal force on the Earth

Socrates
That sounds like your double force nonsense again. With a doubled force the object must be pulled into the middle of the Earth.

Newton
No, no no, this is the clever bit. The ground exerts an equal and opposite reaction to the force that your feet are exerting on it.

Socrates
This again is completely wrong. We cannot evoke the same law twice over. What you just said was that there is a reaction to Gravity and then a reaction to that reaction.
Let me give you my interpretation of what is happening. Objects **can** be stopped in their acceleration by an opposite reaction but it will only happen after a period of time and it may not happen at all. You might, for example, smack into the ground. There would be deformity of your body and of the ground but your motion would be stopped. The force stopping you was stronger than the force keeping you moving and "equal and opposite" has nothing to do with it.
Let me put it another way :

You are pulled to the floor by the force of gravity. You pull back with a much smaller force related to your mass.
You are prevented from going through the floor by the combined forces against the motion of electromagnetism, weak interaction and strong interaction manifest as density and strength of the floor. The strongest object is the one that will deform the least and the weakest object will deform the most. The floor is prevented from being pulled to the centre of the planet by the density and combined forces of the material between the floor and the centre of the planet. If the material of the planet was not dense there would not be such a large mass in the planet for a given size and a subsequent large acceleration due to gravity.

Newton
Well, perhaps I overstated my case. The problem is that there is no such thing as anti-gravity

Socrates
Well, what stopped you going through the floor ..... the forces were acting against gravity were they not ?

Newton

I don't want to talk about gravity any more. The law is true for forces other than gravity. If you push someone away from you when you are swimming in a lake you will be flung backwards as well. I would predict that the same thing would happen in outer space

Socrates

Ah, yes, but the other swimmer does not necessarily push me back unless he is an obnoxious fellow. That is not an example of an equal and opposite force ..... that's an example of a force shared. I would predict that the heavier person would move less than the lighter one.
Let me give you another example of where your law is wrong.
If you exerted a force on a cup to push it along it would not move if there was an equal and opposite reaction. We, of course, know that cups do move so you are wrong.

Newton
Got you now. The equal and opposite reaction occurs because you are pushing on the floor to stop yourself moving backwards.

Socrates
I concur that if I am not moving when exerting a force there must be an equal and opposite force but I was talking about the cup, not myself.

Newton
Well, perhaps my law is only correct for bodies that do not move when a force is applied. How did I make this mistake?

Socrates
By making measurement at equilibrium when objects have stopped moving.

Newton
Well at least you will be able to tell people that I got it wrong. I can't correct it now.

Socrates
I'm afraid I won't be able to. I have been dead for nearly 2,400 years

Newton
Someone soon will put it right.

Socrates
I would not place bets on it. Its still in textbooks at the end of the twentieth century.

Newton
That's terrible

Socrates
Yes it is

What have we learnt from this discourse

Newton's third law is incorrect.

The forces that oppose Gravity are

a) not instantaneous or no movement would have occured
b) work locally but not at a distance
c) not necessarily equal to gravity .... they may be less (air resistance at low velocities) or they may be more (the ground when you smack into it)

Gravity

According to Newton, every mass attracts every other mass with a force that for any two masses is directly proportional to the product of the masses involved and inversely proportional to the square of the distance separating them.

$$\text{Force} = \frac{Gm_1m_2}{d^2}$$

Where the force of gravity between two objects is found by multiplying their masses, dividing by the square of the distance between their centres and then multiplying this result by the constant G :

G approximately = $6.67 \times 10^{-11} N.m^2/kg^2$

(Note that when the measurements are made there is none of this doubling and trebling of the force which the third law would suggest.)

More about gravity later.

Bibliography

Conceptual Physics
seventh edition Paul Gittewitt (Harper Collins)

This is an excellent book and the explanations in it of the Ancients and the Newtonian laws are very clear. Books like this will be classics in a few years time but you do not have to believe every word in it.

Question
What happens to the dimensions after they are used ?
What keeps the mass moving with "momentum" ?
What is time squared?
Why is G only approximate ?

Chapter 3

Space

Human beings are made from star dust. True or false?

There are many myths about the creation of the world but the most accepted **creation** story is that of Genesis. This is the traditional Jewish version as also accepted by Christianity to be found in the Bible in Genesis Chapter 1 verses 1-3.

"In the beginning God created the heaven and the earth. The earth was without form or void and darkness was upon the face of the deep. ..... And God said Let there be light and there was light."

There have been many **scientific** theories about the origins of the universe which have included steady state and constant creation.

But the most accepted theory is the standard Big Bang theory as laid out clearly in the Brief History of Time by Stephen Hawking.
"In 1929, Edwin Hubble made the landmark observation that wherever you look, distant galaxies are moving rapidly away from us. In other words, the universe is expanding."
Hawking then says "This means that at earlier times objects would have been closer together.... about twenty thousand million years ago, when they were all at exactly the same place .... the density of the universe was infinite"

Hawking, by the way does not actually believe this but accepts a modified version in which space and time expand as touching dimensions together and that the sudden expansion of the Universe (known as inflation) occurred after the initial period according to Andrei Linde's version which is known as the chaotic inflationary model. Hawking believes that the Universe is in some way expanding into itself and there is therefore no need to envisage something outside it. This is hard to understand and I do not think that it can be argued logically without including further spatial dimension(s) outside the dimensions of the observable universe but I will explain this in the chapter on extra spatial dimensions.

The infinities are not the only problem with the big bang theopry. The measurements we are able to make seem to show that according to the Big Bang theory the Universe is younger than the age of the oldest star, which is very hard to understand.

Moreover, if all the "forces and energy" were present at the very beginning and if they were symmetrical the forcecs would have cancelled each other out aqnd there wouold have been no expansion. This can only be overcome by assuming that they were not symmetrical in some way.

All of the scientific models next include the formation of the first stars followed by supernovas and the formation of our present earth and sun from the resultant tumult of star dust. This means that everything on Earth is made from the leftover dust from a supernova. You and I are indeed made from star dust that followed a supernova !

But how does the accepted scientific theory coincide with the religious doctrine. Very well considering the age of the text. If you ignore how it is set in motion as presently unknowable (God or chance ?) then the creation sequence goes .... first the heavens, then the earth (initially without form or void), then the light.
The scientific theory goes first the heavens and first stars, next the supernova, then the star dust and out of the formless star dust the Sun and the Earth.

The main difference between the scientific method and religion lies not in the information content but in how you are supposed to deal with it.

In the scientific method a theory is suggested to explain the previous knowledge and make further predictions. If it is shown by observation to be incorrect it may be modified or a new theory which explains the situation more elegantly may be devised to take its place. If two separate but incompatible theories are in circulation the scientist will realise that the theories are incomplete or wrong.

In religion the creed is laid down as being true. The followers are exhorted to have faith and to accept the teachings of the prophet / god. If two parts of the creed are incompatible the follower is told that he must accept both parts of the teaching.

If scientists find that they cannot accept new data because it disagrees with established theory and if they are obliged to believe two or more incompatible theories then the science has been turned into a religion and is no longer a science.

Galaxies

The galaxies are scattered throughout the universe in galactic clusters. No two galaxies look exactly the same but there are three main shapes : spiral galaxies with arms and looking like discs from the side, spheroid galaxies and irregular galaxies. No two galaxies look exactly alike.
 The spiral galaxies are actively producing new stars but the more spheroid galaxies are not. Initially it was thought that they were of different ages ..... the spiral galaxies would then be producing stars because they were younger. But they are now considered to be all the same age and there is no accepted reason for the fact that some are producing new stars and some are not.
The distant galaxies are all moving apart very quickly and the universe is expanding but we do not know why this is the case.

So several questions are raised.

Is the present big bang theory to be treated as science or religion ?
Why are the galaxies moving apart so rapidly ?
What is the universe expanding into ?
Why does the universe continue to expand ?
Why is the universe apparently younger than the oldest star ?
Why are some galaxies producing new stars whilst others are not ?

THERE IS MORE SPACE FOR SPACE LATER IN THE BOOK

CHAPTER 4
TIME

Time is a subject of great debate at the moment. In classical Newtonian, Relativity and Quantum Theories time can flow equally well in two directions. The laws are expressed as closed equations and therefore cannot permit a one way passage of time.

Yet we know that in the real world time does not go backwards. We are born, we live and we die. The seasons come and they go. The clock goes forwards and never backwards.

This is said by some scientists to be something that only really happens due to thermodynamic effects and the build up of entropy in the system. But thermodynamics is only a special case of velocity so why does time not enter into the laws of velocity with an arrow of direction? All those equations **using** time and they never include times most obvious feature .... **it is unidirectional.**

There is also a disparity between our sense of time and the objective measured time. When we are excited by something (our first skiing trip, a roller coaster, a very exciting book) time passes very quickly. This can be associated with a response we know as the "flight and fight" reaction, probably partially mediated by adrenaline. When we are bored and doing repetitive tasks or "nothing at all", time passes very slowly.

This we are told is because it is purely subjective ..... your sense of time and timing is hopelessly inaccurate.
But the sense of timing is amazingly acute in some people ..... witness the Jazz musician or the exponent of Latin American music ..... we know when it is right even though the differences in timing could be only a few milliseconds. Music runs according to very strict rules of timing ..... for example Scott Joplin's ragtime music consists of 4 themes of sixteen 4/4 bars with complicated syncopation in each bar. All the members of a jazz band could play Scott Joplin together whilst falling off a log because his music is very simple compared with the music they normally play!

This is a very developed sense that is usually ignored in textbooks of Biology but is at least noted at the level of daily, monthly and yearly rhythms.
Einstein thought that time was a myth foisted on us by the nature of our brains. How could a myth be so accurate?

What does folk law, legend and common parlance have to say about time? Much wisdom is hidden in the way we use language but also much wisdom is hidden by language.

Common phrases regarding time

Time and tide wait for no man
There's a time and a place for everything
I'll have to make time for it
Its really eating up the time
Time passed so quickly
Give me time and space
Just passing time idly
I was moving so fast (working so hard, enjoying myself so much) the time whizzed by in a flash.
All good times must come to an end.
There is no time like the present
Procrastination is the thief of time

What does this tell us

a) Time moves for us in one direction only
b) Time passes at different speeds depending on what we are doing or thinking
c) But we know that our sense of time and timing can be highly accurate.
d) Any specific time (eg good times) must come to an end
e) Subjectively, we make time, we use time and we eat time or time just passes.
f) We can plan to do things in the future but can only do them in the present (now)

Religions have a lot to say about time but most is related to an eternity elsewhere. We should, however, ignore their thoughts at our peril.

I am not a religious person .... I am a realist. I do not personally believe in the Bible but I consider it to be the writings of profound thinkers. I could be wrong ... it may be more than that or less.

Some phrases in the Bible regarding time are interesting and provide further insight:

There is a time to every purpose Eccl 3.1
Time and chance happeneth to all Eccl 9.11
Endure but for a time Mk 4.17

These are straightforward: If there has to be a time for a purpose this means that if you have an aim you need to make time to achieve it. **Do not reverse the text** : do not say "there is a purpose for every time" which is how most people would interpret the saying and would then not understand that time and chance affect everything. Endurance for a limited period of time is, of course, a well known state of affairs.

More weird and futuristic prophesy :
Time shall be no more Rev 10.6 ...and the most unusual quote of all comes from Daniel 7 verse 25 ..... "and they shall be given into his hand until a time and times and the dividing of time"

These texts suggest that there could be a condition without time and that time may be multiplied or divided. For the latter to occur multiple dimensions of time would be required.

Other religions suggest that time may be cyclical and that we may return in some other form.

What have we learned from Religion

1) For actions (or purpose) to occur time must be made available
2) Time and chance affect everything
3) Circumstances endure only for a time...they do not last for ever !

and the imagination (or revelation, depending on your viewpoint) has added :

1) There could be a state without time (eternity)
2) There could be multiple dimensions of time
3) Time could be cyclical, repeating itself.

Science Fiction has added the possibility of time travel (HG Wells) and also agreed with the possibility of multiple dimensions of time.

Why am I detailing all of this?

The chaos of thought available in well known sayings, the ramblings of weird prophets and the science fiction should all be treated in a similar manner in a scientific thesis. They can be used as a source of new ideas to be examined and tested but they should not be believed.

For a full discourse on the scientific evidence that time goes in only one direction in our universe, I would point the interested reader to "The Arrow of Time" by Coveney and Highfield (Flamingo 1991)

Could it be possible that time could go in only one direction in our universe but that somewhere else could be timeless, or that multiple dimensions of time could exist, or even that time could flow backwards ? Yes it could be so but it would be very difficult for us to explore such a place.

Would it be possible to build the arrow of time into equations without losing the science that we already understand ?

I believe so and will show you how if you read on.

There is more time for time later in the book !

CHAPTER 5

Bar Magnets

Nothing could be simpler than bar magnets, or so we are led to believe.

The reader can now join in with the experimentation.

The theory is that a bar magnet has a north and a south pole. Like poles repel and unlike poles attract.

Take a simple bar magnet and push it towards another bar magnet.
The other magnet will either move away or jump towards the first magnet.
Compress the magnets together against the repulsion and let go of one magnet.
Assuming that they do not stick together (perversely they sometimes do) the loose magnet will move away.
Try with one of the magnets turned round and the magnets will jump together.

Take iron filings and scatter them onto a sheet of paper and put the bar magnet underneath. Lines of force are traced out by the iron filings. These lines are not, however, the real lines of force ..... they are just the demonstration by the iron filings. Try making patterns with two magnets.

The real lines of force are defined in the following way :
"The lines of force of a bar magnet are the imaginary lines traced out by a free North pole from the North Pole of the magnet to the South Pole."

Newton
Well, at least there is nothing wrong with the theory of magnetism. They've even linked it with electricity.

Socrates
You don't think there is anything wrong, eh ? What about the definition of a line of force.

Newton
Seems to be fine. Why, I could have written it myself.

Socrates
That's exactly what is worrying me. It is completely illogical. Firstly you cannot objectively demonstrate an imaginary line.

Newton
Of course I can. I just imagined a line here in space and now I am moving my arm to show you where it is.

Socrates
That is subjectively demonstrating it. **I** cannot demonstrate **your** imaginary line. That is not all. The free north pole they refer to doesn't seem to exist.

Newton
Why not

Socrates
It would be a magnetic monopole and the nearest one could imagine to that is gravity. Magnetic monopoles have never been discovered.
If lines of force can be demonstrated they must exist in some form.

BAR MAGNETS CONTINUED
Making a model of a bar magnet

We could imagine attempting to make a model of a bar magnet out of firstly wood and springs then secondly wood and elastic.

Place the spring between two pieces of wood. Compress the pieces of wood together and then let go. The wood will move apart.

Take away the spring and relace it with elastic tied around the wood. Pull the two pieces of wood apart and let go. The wood will move together.

Apart from the different materials involved does this mimic two bar magnets ?

For the first part the answer is yes : the force pushing the two magnets apart after you pushed them together was the stored energy from your own compression of the magnets together, just as it was with the pieces of wood and the spring.

But the force pulling the magnets together was quite different from the elastic pulling the wood together. With the magnets as you pushed the first magnet towards the second, the second magnet suddenly moved towards the first magnet and stuck to it. When you pulled them apart the strength of pull was greatest when they were nearer together, becoming weaker as you moved them further away from each other. With the pieces of wood and the elastic, the pull was least when they were closest together becoming greater as you pulled them apart until you let go. The pieces of wood then moved back together again using the potential energy that you had stored in the elastic.

The difference is that the energy used in both spring and elastic was your own as was the energy used pushing the repelling surfaces of the magnet together (the "like poles"). But the energy making the magnets pull towards each other was generated by the magnets. So the two aspects of the magnetic force are not symmetrical ..... one releases energy but the other uses your energy of deformation of the lines of force.

Newton
That's a lot more complicated than I expected
Socrates
Wait till you read about fridge magnets in the second section of this book

Newton
What's a fridge ?

Chapter 6

Quantum Theory

Quantum Theory has been developed by some of the best minds this century and the author of this book does not pretend to understand all of the mathematics involved or all of the experiments, both real and thought experiments, that have been carried out. What I have done here is to try to find the essential concepts behind the theory. To the quantum theorists I would say : I am sorry if it is too simplistic but it is how I understand the situation.

Quantum theory consists of two components

1) Emission and absorption of radiation by a body can only occur in finite packets of energy which are equal to h times the frequency (where h is the original Planck's constant). This was discovered by Max Planck at the trun of the century.

2) A system of probabilities which have been measured or predicted regarding particles. The statistics permit us to know the probability of the particles being in a certain place and having certain properties but we cannot know the full answer . The probabilities are known as the Wavefunction.

The first part of the theory, that things come in small equal-sized packets for a given wavelength, is not difficult to accept although it came as a big surprise to the Victorians .

The second section is very difficult. The most peculiar aspects of the theory include

a) Wave-particle duality: Particles can have two related properties but you cannot see both at the same time. When you show the wave nature you cannot see the particle and vice versa. When you visualise interference due to a wave you cannot see the particle. If the particle is detected you cannot see the wave.
There are many other dual or complementary properties in which the better you know one property the worse you know the other within a range of probabilities

b)The probability or chance that the particle may be here or in a remote part of the universe.

c) The problem of Einstein,Podolsky, Rosen theory and entangled pairs. If you change a factor of a particle here even if the other particle is across the other side of the Galaxy it would theoretically also change.

d) Faster than light quantum tunnelling. A photon (electron etc.) may mysteriously have a small chance of apparently passing through an otherwise impermeable barrier that would normally absorb or reflect it. In multiple experiments where a large number of photons were directed at a barrier a small number of the photons passed through the barrier **faster than the speed of light in a vacuum.** This final fact is so astounding that scientists in this area have been unable to get their work taken seriously for the last forty years.

Newton

If these effects are so bizarre how come they are not obvious to us in the real world?

Socrates

You can't argue with this, it's established fact. You're only allowed to argue with Goddard's theories.

Newton
I thought you insisted on logic in all things.

Socrates

OK, OK
The reason that you cannot observe these bizarre phenomena is that they occur at a very small level. You are too large for the effects to be noticeable. The larger you are and the greater the number of particles you contain, the smaller is the chance that the effects can make an obvious impact

Newton

But theoretically there is a chance that I could be here or in Alpha Centauri, and that I could move Faster Than Light between the two ?

Socrates

In theory there is a very, very, very remote chance of that happening. That is the way with statistics like this. The only way in which you can be certain that it has not happened is to know where you, or the particle, are right now. You cannot know that it will not happen in the future.

Newton
So I can ignore the theory since it can only remotely affect me.

Socrates
You cannot ignore it although many people do. All they have taken into their vocabulary is the term uncertainty ... it may of course be one of the reasons that so many people seem uncertain of the future today. Its almost become a religion.. You forget about it but just believe in it.

Newton

Why should I remember it at all

Socrates
Because it comes into everyday life because of its importance in electronics.
If you were a radiologist like Dr Goddard you would be aware of Quantum mottle on X-rays and quantum effects on MRI scanning sequences.

Newton
This theory seems to be full of imponderables. I like that. I think I will call it Newton's law of imponderables.

Socrates
I'm sorry old boy. Heisenberg got there first. He's called it his Principle of Uncertainty.

Newton
I'm not so pleased with the idea now. All he is doing is giving it a fancy name. That does not solve anything.

Socrates
Who's calling the kettle black. Look at you with momentum. Galileo discovers that once having started with a velocity an object will keep moving with the same speed in a straight line unless a force acts on it. You come along as the greatest man in science and give it a name "Momentum". You don't explain it like Aristotle tried to ..... no, no ,no . You name it and then everybody is happy and ignores it for hundreds of years.

But I agree that the Principle of Uncertainty is no principle at all. Its just a name for an observation. Some clever people have tried to find the reason for the uncertainty and they have come up with a variety of possibilities..... some sensible and other ideas just as bizarre as Quantum Theory itself.

To name but a few

Hidden variables in the photon
Hidden variables in the source of photons
Multiple worlds
The effect of our own consciousness on the experiment

Unfortunately no single answer satisfies all the criteria. In particular hidden variables in the photon and source of photons or other particles cannot solve the distant problems such as quantum tunnelling. Many scientists have given up and just accept it as an inevitable fact of life. If they use quantum theory at all they use it like an adding machine.

Newton

Well, that's OK then. It works so lets forget about it.

Socrates

No, that is not alright.

I'll give you an example that I have heard Goddard quote from his medical experience.

A patient went to his doctor and complained "Doctor ... I don't know what is wrong with me. I think I must have some scarring on my lungs but I don't know why... I can't breathe like I used to and I'm short of breath"

The GP did some blood tests and sent the patient for respiratory function studies and a chest X-ray. The respiratory physician ... a wise fellow.... sent the patient for CT scans. Dr. Goddard saw the CT scans and made a diagnosis of Cryptogenic Fibrosing Alveolitis. Dr. Goddard and the physician discussed the patient and could find no hidden cause for the condition.

The patient went back to the GP who told him that he had Cryptogenic Fibrosing Alveolitis. A course of treatment was started but it was not expected that it would make much difference.
The patient was happy because there was a diagnosis. But the GP, the respiratory physician and Dr. Goddard were unhappy.

Newton

Why were they unhappy ?

Socrates
Well, that's easy to answer. You see Crypto means "hidden", Genic means "cause", Fibrosing means "Scarring" and Alveolitis means " Disease of Lung".
So when we work it all out, all he is doing is giving it a fancy name and the patient knew all along that he had scarring of unknown cause. That does not solve anything.

No doctor would be happy with that and in medicine they are always looking for the hidden causes. The most difficult diseases are those that have several contributory factors and may result in disease only when the causes act together.

Questions

Should we be satisfied with the Uncertainty Principle ?

Could there be multiple causes of the statistical probabilities that result in the Uncertainty ? Can we identify any of them ?

Bibliography
For a good introduction to classical Quantum Mechanics try *The mystery of the Quantum World* by Euan Squires, Institute of Physics Publishing, Bristol

CHAPTER 7
Special Theory of Relativity

Einstein completely altered the way in which we looked at the world. He showed that there was an interrelationship between mass, space, time and energy and this is given in the famous equation:

$$E = mc^2$$

E is energy, m is mass and c^2 is the velocity of light squared.
He was demonstrating the equivalence of mass and energy.

This is translated into the dimensions

$$\text{Energy} = \text{Mass} \times \frac{L \times L}{T \times T}$$

or energy $= \frac{ML^2}{T^2} = ML^2T^{-2}$

Einstein argued that the maximum velocity that could be achieved was the velocity of light in a vacuum. Since this was the maximum velocity even if two light beams were moving away from each other they would not be able to exceed light speed and time itself must alter to cope with this. This would mean that time is relative and depended on the velocity of the observer. If you were an observer in a space ship, as you moved faster, time would dilate for you. You would experience no difference according to your watch, which would be moving with you. But an observer on Earth would still be affected by the velocity of movement of the Earth (and the Force of Gravity) and would be experiencing time at a standard rate. He would age more than you because time had not dilated for him.
There are other consequences of special relativity:
As you move faster your mass appears to increase
As you move faster your length in the direction of the movement appears to decrease to the observer on Earth whilst staying normal to you (Lorentz-Fitzgerald contraction).

Einstein said that we should consider it as one continuum called space-time.

Finally, Einstein said that if you moved faster than light you would be moving back in time.

These findings were amazing to a public which had access to excellent clocks and watches and from the time of Newton had thought that time moved at a steady pace.

The theory was called the special theory of relativity because it specifically referred to uniformly moving objects (velocities) but did not refer to acceleration.

Newton
I've heard about this guy. Do I have to believe any of it?

Socrates
I'm afraid that you do. You have no choice but believe some of it because experimentation has proven it to be true.

You do not have to accept that the speed of light in a vacuum is the fastest possible velocity since we know that "Faster than light" speeds can be achieved due to quantum tunnelling and you do not have to accept that moving faster than light would make you move back in time.....that has never been experimentally confirmed and there is no evidence that the quantum tunnelling made the particles move back in time.

Newton
What do I have to accept. I don't want to accept any of it, if it does not agree with my laws of movement.

Socrates
Well, we have already removed your third law

Newton (embarassed)
Apart from that one

Socrates
It seems that your first two laws are correct at slow velocities in a fixed gravitational field.

Newton
But does experimentation agree with Einstein

Socrates
OK. here is an experiment that has been done. Take two atomic clocks and put one on a fast aeroplane such as Concorde. Send that one around the world but keep the other one stationary at home. When the clocks are back next to each other the one that went in the plane is running just behind the other due to time dilation.

Newton
I think the aeroplane, noisy contraption, has interfered with the clock's mechanism and it is broken.

Socrates
No, the world-hopping clock is now sitting next to the home-loving one and it is running at exactly the same rate as the other one but is slightly behind it.

Newton
Do any other parts of his theory work

Socrates
Yes they do.
If you accelerate particles close to light speed they appear to become more massive.

Newton
Perhaps it is not important if I am here on Earth

Socrates

Here's an interesting fact . There are particles that are produced due to the cosmic radiation hitting Earth. They should not reach us on the surface because of their short half lives. But they **do reach the surface** because time dilates for them due to their immense velocity.

Newton
I cannot explain that.

Socrates
Well, now that faster than light velocity has been confirmed nor could Einstein if he were here now.

Question
Why does time dilate as you increase your velocity ?

CHAPTER 8

General Relativity

Einstein went on from his Special Theory of Relativity, which referred to objects at constant velocities, to create a General Theory of Relativity which analysed objects that were accelerating.
In this theory he stated that a mass would result in the bending of the three spatial dimensions in the fourth dimension of time which should be considered as part of a space/time continuum. This, he concluded, was the cause of gravity and would result in the path of a photon of light being bent even though it was massless. Objects which had mass would also bend space and would thus be more affected by gravity.

Einstein's predictions regarding the apparent bending of light by gravity were demonstrated to be true by viewing an eclipse of the Sun and showing that light from the stars did indeed bend around the Sun.

Einstein could find no difference between acceleration from any cause and gravity.

Newton
Thats nonsense, he's spoilt my simple concept of gravity as a force.

Socrates
You only named it, you did no more

Newton
Well, he is now saying that it is exactly the same as acceleration. It can't be. Acceleration does not obey the inverse square law

Socrates
Yes but its action is indistinguishable from that of gravity

Question
Why do you think that mass bends space?
Is it true that gravity is bent space (as theorised by Einstein) or is it possible that space is bent by gravity ?
What is gravity ?
How can gravity and acceleration be indistinguishable in this theory but qualitatively different in real life ?

INTERLUDE

THE HORSE'S MOUTH

What do we observe when we see a car accelerating or an arrow flying through the air. Do we see energy being expended and force being used ? Do we see an example of a body with momentum ?

Yes, is the unanimous reply.

We think that we do because such concepts are firmly rooted in our subconscious and in all of our language. It is almost impossible to think without using such words.

But we do not see energy, force, or momentum. We deduce they are there because the car is accelerating and the arrow continuing to move (albeit with a slight deceleration).

What we actually see is the car moving faster as time passes. We see change in velocity per second. Put another way, we see the car using dimensions as it accelerates. They are LT^{-2} (acceleration) by L, as it moves along.

We see the arrow moving with what appears to be almost a constant velocity. We are observing the use of dimensions of length per time. $\frac{L}{T}$

If we believe that the strongest evidence obtainable is that which is directly observed, **we have to accept that the observation that spatial and time dimensions are used is stronger evidence than the concept that energy, force and momentum have been used.** The latter concepts are just names to explain the situation.

Newton

But you cannot use spatial and time dimensions without using energy, force and momentum otherwise how did the movement happen ?

Socrates
Really, is that so ? You cannot imagine another concept to explain what you are seeing ?

Newton
Well, I can certainly imagine another concept. My imagination is not that stultified. But energy, force and momentum are the best concept. You can't just add and subtract, multiply and divide spatial dimensions.

Socrates
That is the one thing that you can do otherwise there would be no point in having a measuring stick. We rely on being able to do that for any measurement. Even time

is usually measured against a spatial marker : the shadow of a sundial, the movement of the sun and moon, the movement of the hands on a clock.

Newton
You are just arguing about words.

Socrates
No, I am telling you about observations. You must believe what you see

Newton
So you think that spatial dimensions are just added at whim.

Socrates
No I do not. I accept that there is cause and effect but do not accept that the standard concepts of energy, force and momentum are necessary.
Dr. Goddard will present a completely logical argument for one alternative. There could be many other possible explanations but they are not pursued in this book.

Newton
But what had the horse's mouth got to do with it ?

Part 2

Goddard's Conceptual Grand Unifying Theory

CHAPTER 9

INTRODUCTION TO PART TWO

This half of the book tries to answer the questions set in the first half of the book with some logical speculation. Throughout this section the suggestions made are not intended to be taken as a final utterance or theory. They are what they say they are ..... a logical model to explain what we see. The ideas should be taken as fertile ground for physics research and should not be treated as anything else.

The biggest problem with science seems to be that we have named a lot of things and then measured them but we have provided no explanation as to how they work. This was acceptable when the ideas such as momentum, gravity, energy, mass and uncertainty were first muted because they permitted an orderly advance in knowledge. But this cannot be considered ideal and we should look for the hidden causes.

In some places several possibilities would fit into the logical speculation. They have therefore been detailed.

The achievements of physicists, mathematicians, cosmologists and sciencists generally this century have been outstanding and too numerous to recall.

But why have there been problems with understanding fundamental factors like time and space, energy and work, momentum, mass and velocity ? Why does it all seem to be so complicated ?

There are several salient reasons why most people do not think hard about these problems :

1) They **are** complicated
2) Many intelligent creative people just cannot be bothered with something so far removed from their usual experience
3) Other people think that the physicists have it all worked out : otherwise how did they manage to get men on the moon ? (answer : the fudge factor)
4) Creative and chaotic people can succeed very well in the arts but find the discipline of science too difficult. They do not understand what science is trying to do and they find the necessary "fact grabbing" too irksome. Those who are good at seeing the concepts behind the facts and laws find science very confusing because "it does not seem to make sense".

Why are scientists so split up into different disciplines ? Why do different branches of science find the other branches incomprehensible and sometimes derisory ? This, I believe, is inherent in the way that science is taught and in the personalities of the people who choose a career in science.
There are major problems in most sciences but physics has particular problems. To obtain a degree in Physics the candidate would have to study Classical Newtonian physics, quantum mechanics and relativity. Since they are not all

compatible and since there are definite areas of conflict he would have to remember the facts rather than comprehending them and this can induce a state of acceptance and belief, rather than questioning and scepticism
(the outlook that a scientist should have).
Let us point out a few problems in physics:

Nothing can travel faster than the speed of light (Einstein)
But quantum tunnelling theoretically permits instantaneous transmission of light (known theoretically since the 1930s and shown experimentally since the 1950s)

Space should be very curved (Einstein)
But space is very flat (Linde)

The big bang theory states that everything started with a very small object which contained all the mass and energy of the universe in an infinitesimally small space and then expanded in an inflationary way without any loss of content
But if this were so it is very hard to conceive of a situation that did not start with an infinitely dense and hot object, which most would consider as absurd.

Time is considered to be theoretically reversible in theories of Newton, Relativity and Quantum mechanics.
The experience of time is that it is not reversible.

There are more specific problems including the fact that when applying present theories there is only 10% of the predicted mass in the Universe, the Universe appears to be younger than the oldest stars, the sun produces too few neutrinoes etc, etc. The recently discovered small mass of a neutrino assists in solving some of these problems but makes others much worse (such as the number of neutrinoes that should come from the sun).

Why should there be all these problems in modern physics?

Physics is supposed to be the underpinning of all the other sciences yet theoretical physics is ignored by all the other sciences. Engineers have their own mathematics, mathematicians cannot understand the logic of theoretical physicists etc. etc.

I believe that there are three fundamental problems:

1) Naming an object or situation is too often considered to be the same as understanding it
2) Measurements are made at equilibrium and concepts are expressed in equations. But in reality situations are usually dynamic and are asymmetrical with reactions occurring in one direction only or imperfectly passing backwards into the other direction. There is an inbuilt arrow of time . Chaos theory is ignored by physicists at their peril since the Universe, consisting of countless billions of interacting particles must be a chaotic system. Even situations that are in theory perfectly described by mathematics other than Chaos theory, are in practice chaotic due to exceedingly minor changes at the molecular, atomic or quantum level.

3) The belief in contradictory theories that set in as a student is difficult to shake later.

This part of the book attempts to shake that belief by showing a new way of looking at familiar data.

In many cases the new theories are just the old ones turned round in order to try and find the hidden reasons. In such situations the mathematics would be unaltered.
In other places the theories are radically new but follow logically from previously accepted information.
In other cases the theories are based on experience of similar situations in other sciences such as chemistry and medicine.
In a few places the theories are logical but even more speculative and in such situations I have indicated the problems.

In the end I believe that I have produced a theory that could represent the basis of a Grand Unifying Theory but in addition may show the way to the inclusion of other branches of knowledge.

This may not be the correct theory (it probably is not) but it deserves an airing and scientists should consider it. .
At the very end you could say that all the author has done is twist science around and give it new names. That can be very necessary because of the emotional illogical baggage that goes with a name. You do not have to believe anything in this book if you do not wish to do so ...... but if you are a scientist you should try to find a logical reason for not doing so and you should try to test the theory against its predictions.

Now is the time for you to read Inventing Reality by Bruce Gregory if you have not already done so. This is a very interesting book with a lot of good reasons for my book being more correct than previous theories. At the time of devising this theory I had not read Gregory but I have now done so and consider it to be a major insight. The conclusion is almost totally correct (that reality is a function of langauge) because so much of reality is hidden behind the words.

It is not completely correct because there is a reality out there as anyone but a solipsist can tell you. Because of this difficulty I have had to invent a lot of new words to explain what I think is happening.

If this second section stimulates thought it may well have done its job

Please read on.

CHAPTER 10
Elementary rules of geometry for 4,5 and 6 spatial dimensions and a law of duality, complementarity and distance probability

Paul R Goddard and
Jeremy P G Goddard .

This crucial part of the theory was worked out in conjunction with my son, Jeremy.

Introduction

Rules of geometry apply to the three spatial dimensions and the progression from one to two and then to three dimensions permits the establishment of laws that can then be applied to four, five, six or more dimensions without resort to complicated mathematics. One can establish how a one dimensional observer in a one dimensional world would respond to a two dimensional object or situation. Similarly, the way in which a two dimensional observer would respond to a three dimensional object or situation can also be established. From extrapolation it is possible to determine how a three dimensional observer would respond to a four dimensional object or situation and the result is a law of duality and uncertainty which corresponds precisely with Heisenberg's Principle of Uncertainty. If time is accepted as one dimension, for a description of the observed space-time continuum four spatial and one time dimension would be required.

Dimensional Rules

Zero spatial dimensions

In zero dimensions there is only a single point

In one spatial dimension

There can be multiple points on a line but two separated points on a line cannot meet. The ends of the dimension cannot meet therefore a finite never-ending line is not possible. Points (zero-dimensional) can be used to define an "inside" and an "outside" and the inside cannot meet the outside.

Figure 1

Points A,B,C and D cannot meet.
Between B and C could be defined as "inside" and cannot meet any "outside" point between A and B or C and D.

A zero-dimensional point can be passed over a stationary point instantaneously but will take a measurable time to move between two points on the one dimension. A movable line cannot move through an impenetrable point or line.

In two spatial dimensions
Any two points on a line can meet as shown for B and C in figure 2a.

The ends of a line can be joined up to produce a never-ending but finite line. Shapes such as a circle, square, triangle etc. can be drawn using a one dimensional line and will define an inside and an outside (E and F in figure 2b) which can never meet in the two dimensional plane.

In two dimensions parallel straight lines (parallel single dimensions) cannot meet and separated points on a two-dimensional plane cannot meet.

A one dimensional line can be curled up very small.

When curled up the differences between them include their touching points where they cross over and their increased density when curled up tight.

A movable line can be passed over a stationary line instantaneously.
A never-ending but finite "ladder" is impossible.

Figure 2a

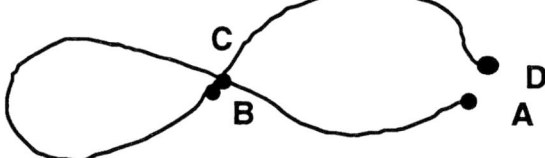

A point on a line can meet any other point in two dimensions

Figure 2b
The inside E can be delineated by a single dimension and will be separated from the outside F such that they will never meet.

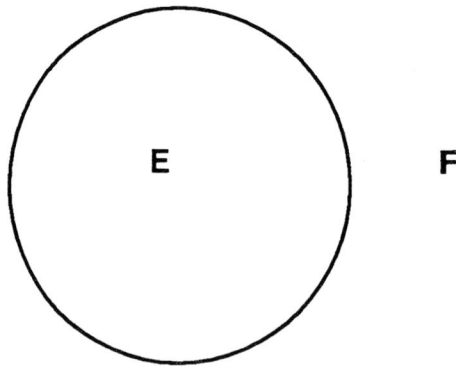

Figure 2c

Parallel lines will never meet in two dimensions

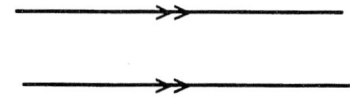

A movable line can go past an impenetrable stationary point on a line by crossing over lines as at BC in figure 2a.

In two dimensions a movable area cannot pass through or over an impenetrable point, line or area but can pass around it.

In three spatial dimensions

In three dimensions any two points on a two dimensional structure can meet whilst remaining separated in the original two dimensions. This can be proven by taking a piece of paper and marking two points on it and then bending the paper such that the two points meet. Thus the inside of a circle can meet the outside. On unfolding the paper it is clear that the two dimensional plane has remained intact and the points on the plane are still separated in the two dimensions.

Parallel straight lines can meet. The parallel lines can be drawn on a flat piece of paper and then the paper bent in a third dimension such that the two lines meet.

A two dimensional plane can be curled up very small in three dimensions.

A never ending "ladder" can be constructed (Figure 3a).

A three dimensional object, eg. a sphere or a cube, can be constructed with an outside and an inside separated by two dimensional planes. The outside and inside will never meet if the sphere or cube remain intact.

Separated points on a three dimensional object, eg. a sphere, cannot meet if the object is to remain intact in the three dimensions.

Parallel flat two-dimensional planes can never meet (figure 3b).

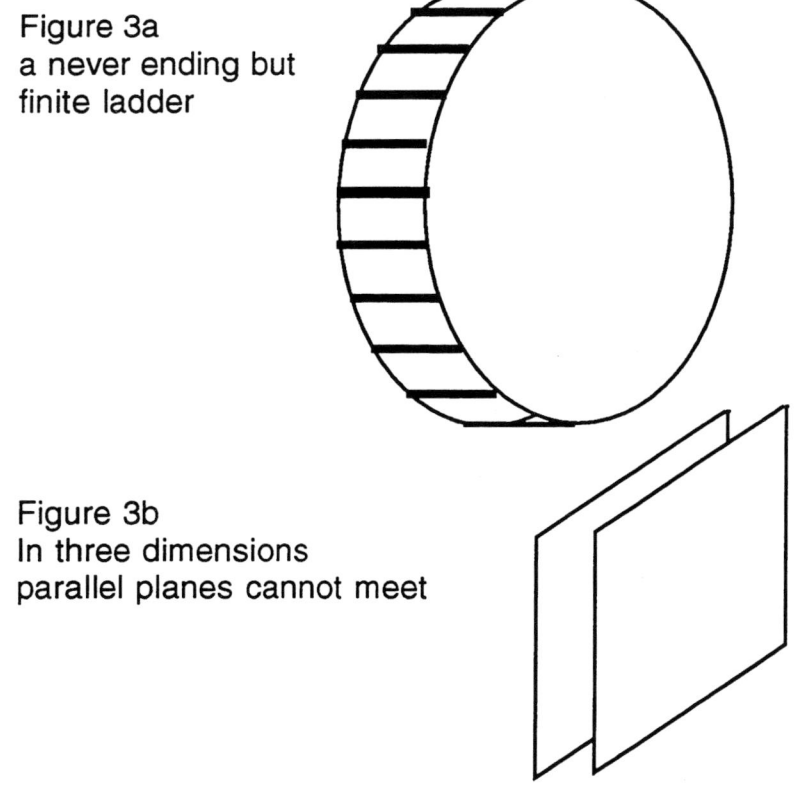

Figure 3a
a never ending but finite ladder

Figure 3b
In three dimensions parallel planes cannot meet

Rules for further dimensions

We can extrapolate from these observations.

In each case when we moved to one more dimension points separated in the former dimensions were able to meet.

Thus in four spatial dimensions any two points in the three dimensions would be able to meet without altering their relationship in the three dimensions. Thus the north and south pole of a globe could meet whilst the globe remained intact in 3 dimensions. A never-ending but finite staircase could be constructed, as could many other objects not possible in three dimensions.

In each case an "inside" and an "outside" (defined by joining the ends of examples of the "number of dimensions minus one") could be made to meet if a further dimension was added.

Thus in four spatial dimensions the inside and outside of a three dimensional object could meet. Thus the inside and outside of a sphere could meet in four dimensions whilst remaining intact in three dimensions.

In each case parallel examples of "the number of dimensions minus one" would never meet (separated points in one dimension, parallel lines in two dimensions, parallel planes in three dimensions).

Thus in four dimensions parallel volumes (three dimensional manifolds) would never meet but parallel flat planes would meet whilst remaining parallel in three dimensions.

In each case it was possible to fold up or coil up an example of "the number of dimensions minus one" into a very small space.

Thus in four spatial dimensions a three dimensional structure could be folded up into a very small space without being altered in its relationships in the three dimensions.

In each dimension a movable example of "dimensions minus one (D-1)" can be passed over a fixed D-1 in one go (point over point in one dimension, line over line in two, area over area in three).

Thus in four spatial dimensions a movable 3D volume could be passed over a fixed 3D volume, over an area, over a line or point instantaneously.

So we have concluded that in four dimensions

Any two points in the three dimensions would be able to meet without altering their relationship in the three dimensions.

The north and south pole of a globe could meet whilst the globe remained intact in 3 dimensions. A never-ending but finite staircase could be constructed .

The inside and outside of a three dimensional object could meet. Thus the inside and outside of a sphere could meet in four dimensions whilst remaining intact in three dimensions.

Parallel flat planes would meet whilst remaining parallel in three dimensions.
Parallel volumes would never meet.

A movable 3 dimensional volume could be passed over a fixed 3D volume, over an area, over a line or point instantaneously.

Five, six or more dimensions

Similar arguments can be used to determine the rules for 5,6 or any number of dimensions.

Thus in five dimensions parallel volumes (three dimensional manifolds) can meet whilst remaining separate in the four dimensions.
Parallel 4D manifolds cannot meet
A four dimensional object can be folded up very small .

In six dimensions parallel 4D manifolds can meet.
Parallel 5D objects (manifolds) cannot meet.
A 5D object can be folded up very small........ and so on.

A zero dimensional observer responding to a one dimensional object

A zero dimensional observer could experience a one dimensional coloured object passing through it as schematically depicted in figure 4

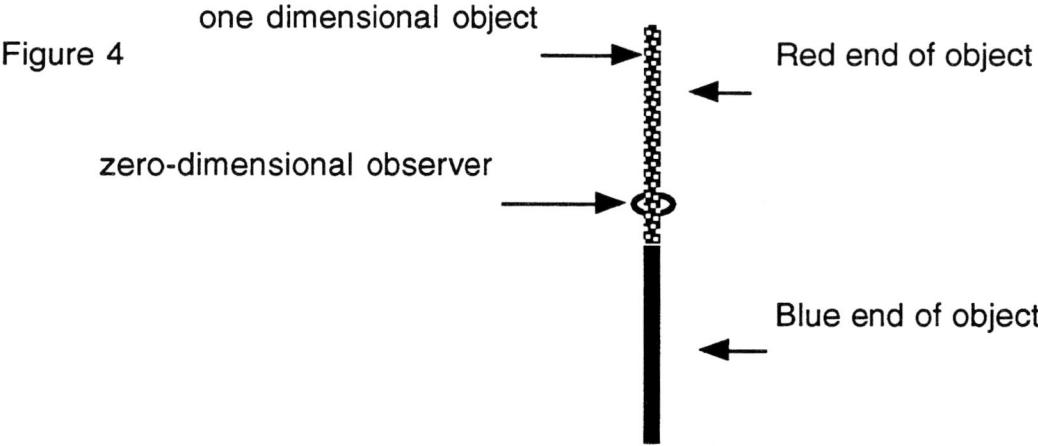

Figure 4

The observer would not be able to experience both colours at the same time but would decide that when it experienced the object it was not permitted to know both colours together. If the object was passed through the observer and out the other side the observer would know that if the blue was experienced first red would follow and vice versa.

There may, however, be three objects (Red/Blue, Red/Green, Blue/Green) to be passed through at random. When it experienced the first colour the observer would, after accumulated experience, realise that there was a probability governing the second colour.

The observer may then formulate a law of duality or complementarity : "When experiencing this peculiar object, I am not permitted to know both colours at the same time. If I first experience Blue there is a 50% probability that I will next experience Red".

A one-dimensional observer responding to a two dimensional object or situation

If a one dimensional observer were to meet a two dimensional object he would only be able to observe the part of the object in his one dimension

Figure 5a
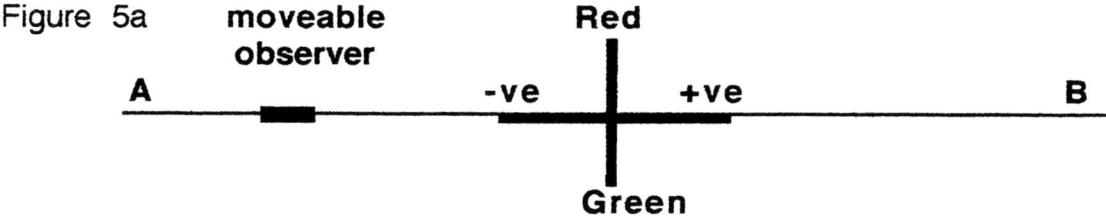

Figure 5b
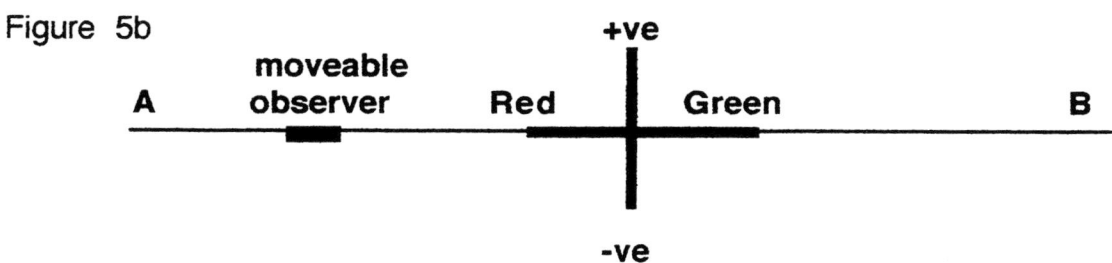

In figure 5a he would be able to say that the positive end pointed towards B and the negative end towards A but he would not know about the red and green limbs. In figure 5b he would know the direction of the red and green limbs but would not know about the positive and negative limbs. He would initially think that they were different objects. However, after considerable observation he would notice that they appeared in the same place and he would create a law of duality ... "The object has both charge and colour. If I know the direction of the charge (+ ve or -ve) I cannot also know the direction of the colour. If I know the direction of the colour I cannot also know the direction of the charge ". The observer may however be able to compute a probability as to which direction the charge will be in (in this case 50%).

A one dimensional observer may observe the passage of a small uni-dimensional object across a crossed line (figure 6).

Figure 6
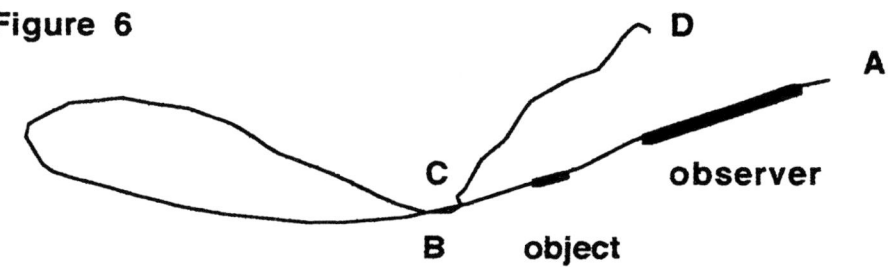

Sometimes the object will pass around the loop in the same way as the larger observer is obliged to do. At other times the object will pass from B to C with no apparent passage of time. This will be incomprehensible to the observer who will

decide that it depends on the small size of the object and will say that there is a probability that the object is both at B and C. The observer will say that it is possible to know where the object is but not where its going or where its going but not where it is. It will also surmise that the probability that the object is in more than one place at the same time becomes smaller as the object increases in size.

A two-dimensional observer responding to a three dimensional object or situation

If a two dimensional observer were to observe a three-dimensional object he would only see the part of the object that was in the two -dimensions (figure 7).

FIGURE 7

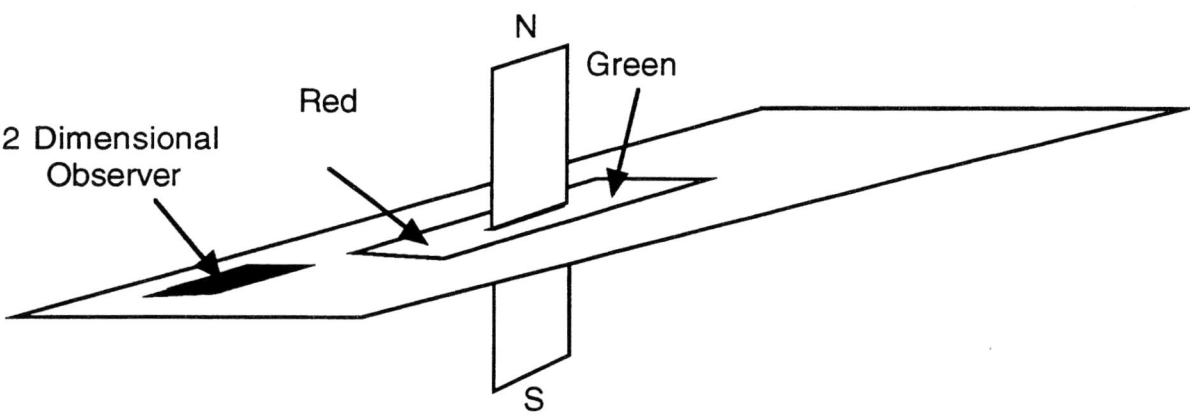

If he observed for a while he would discover that the object had polarity and colour but that he was not able to know the direction of both at the same time.
He may ascribe this as a principle...."The object has both polarity and colour. If I know the direction of the polarity (N or S) I cannot also know the direction of the colour. If I know the direction of the colour I cannot also know the direction of the polarity". Again he may be able to compute a probability. There may be multiple similar but slightly different objects (say with colour and polarity, colour and charge) and the probability calculation would again be calculable.

The 2D observer may be on a folded plane and not know it. Then any part of the plane may meet any other part and depending on size, two dimensional objects may pass from one part of the plane to another with no apparent passage of time. Again the observer may enshrine this in a law of probability and uncertainty which permits such things to happen but he would not understand it.

Consider also an observer looking at the lines of force around an imaginary two dimensional magnet. When the magnet was turned round the lines of force, which were spreading out indefinitely into the surrounding two D world, would also turn round and a wave may be produced. He may consider "How it can be possible that the lines of force that spread out into my entire world can all be moved round so quickly. ..... even the lines of force at the very furthest distance in my two D world?" But a 3 dimensional observer could see that a 2D magnet with 2D lines of force could be folded such that the lines of force do not have to move rapidly in 3D even though they appear to do so in 2 dimensions.

Now let us consider what would happen to a three dimensional observer in a 4 spatial dimensional world.

4 dimensional objects (manifolds) could have linked properties such that only one property can be seen in the three dimensional world when the other property is projecting into the fourth spatial dimension and therefore not observable.
Thus an object could have charge and polarity, position and momentum, particle and wave properties but only one of the two linked properties could be observed at a time in three dimensions.

A three dimensional observer may observe the passage of a small object between two points with no apparent passage of time due to the complex folding of space such that two points were touching . He may find this very puzzling since in other circumstances there may be a maximum velocity (c).

The spatially four dimensional observer would understand all of these peculiar phenomena. Duality would be due to one of the properties being turned into the fourth spatial dimension . Objects could pass from one point in three dimensions to another point in three dimensions because the points are touching in the fourth dimension. The 3D observer would find this unusual because he himself, being large, may not pass through such touching points if they are small.

Discussion

The postulate of an extra spatial dimension may be used as a model to explain some of the phenomena that have been observed in the real world at the level of quantum mechanics. They are included in Heisenberg's Principle of Uncertainty: complementarity and duality. However, if the above postulate of an extra spatial dimension is to be used as a model it should be true that complementarity (that an object may have complementary features that are not always detectable) would be more important than Heisenberg's Principle of Uncertainty. This is indeed considered by some to be the case (The Duality of Matter and Light , Berthold-Georg Englert et al Scientific American, Dec 1994 pp 56-61).

In addition there should be circumstances in which objects can move distances apparently without passage of time and this does appear to be the case with quantum tunnelling (Faster than the speed of light, Julian Brown, New Scientist 146, 1.4.95 pp 26-30). Moreover there are "entangled" pairs of particles whose destinies are linked even if they are far apart as described by the Einstein-Podolsky-Rosen theorem (New Scientist 3rd April 1993 p 12) and they could be explained in the proposed model by linkage in the extra spatial dimension.

The addition of an extra spatial dimension does not explain why the observed effects occur at such a small scale. If the extra dimension were exactly like the other three spatial dimensions one would expect that the observed phenomena described above would occur at all sizes ..... thus whole galaxies might move through the extra dimension with no passage of time and we might be able to move ourselves through the extra dimension. This does not appear to happen in reality and the peculiarly bizarre phenomena of quantum mechanics seem to affect only very small objects. This is explainable if the extra spatial dimension only touches the other three dimensions at very small points or along only one or two of the dimensions. In such circumstances only small objects or structures of two or less dimensions would be able to pass through and larger objects could pass through only if all of the small objects they were made of were able to move together through the dimension. This would have a very low probability approaching, but not reaching, zero as the the object increased in size.

It is not entirely necessary to invoke a fifth dimension, if some aspects of time can be used as a spatial dimension but it is easier to consider it as a different spatial dimension.
If a fourth spatial dimension is present and is part of the cause of the phenomena observable at the quantum level it may be possible to construct an experiment to demonstrate it.

This chapter is not intended as proof of an extra spatial dimension but rather as an argument to demonstrate that the postulate is a logical and viable model which may be the best explanation available for some of the peculiarities observed at the quantum level.

The method of derivation described in this article is rigorously logical and is therefore a true mathematical description . Moreover, using the simple method of extrapolation described, readers of this chapter will be able to devise more laws for the extra spatial dimensions and for one dimensional observers meeting three dimensional objects (and so on).

We have decided to call the extra spatial dimension(s) Dq for quantum dimension since it is observable at the quantum level and it will be described as Dq for the rest of the book.

Conclusion

Goddards' Law
If an n dimensional observer met several n+1 dimensional objects each with dual properties, he would be able to devise a law of duality.

If an n dimensional observer's n dimensions were folding and touching in n+1 dimensions, on observing an n dimensional object he may see it move long distances in n dimensions without any passage of time as it moved across the n dimensions which were curved and touching in n+1 dimensions.

CHAPTER 11
The important whimper
The parasitic universe and the self-replicating particle

Now that we have introduced the concept of Dq, I would like to introduce a few more concepts. These have already been seen in a different or analogous way elsewhere but will be introduced here under different names.

The big bang

In the section on space we posed several difficult questions, some of which are rhetorical. Perhaps the most important question was

Is there a viable alternative to the Big Bang that does not include any Infinities ?

The answer has to be yes.

One logical alternative is a parasitic universe.

That does not sound very nice but it is the best description. I prefer the self-replicating universe because it sounds more appealing but it is less accurate since it is the particles that replicate in this theory not the universe

Andre Linde has put forward a cogent argument for the universe having developed in chaos as one of multiple universes. By an evolutionary process eventually one universe with the correct dimensions and properties would exist. He concludes in a major review in Scientific American (Linde: Sci Am. Nov 94) that only a Fractal model of the inflationary period of the Universe could result in that which we see now.
With much of this I agree and Linde's work has been an inspiration to me to keep going when other people told me I must be wrong (or so they believed).

In the parasitic universe theory the universe would have started in the chaos with one particle and expanded by **addition of dimensions and particles from a surrounding chaos of dimensions** by acting as an attracting agent. The chaos will be called Cahotage since chaos has many different meanings to each person: I know what I mean by it, but we do not wish to confuse the issue.

In the past the attracting agents have been called force. In this theory force will be called an **activator** since it allows "energy" to be used or more correctly work to be done.

The edge of the universe would continue to expand by addition of dimensions of space and time and the dimensions incorporated from Cahotage would be incorporated into our Universe and would be manifest as entropy. The dimensions thus incorporated would add to the 3D size of the universe and to the size of Dq. This could explain why some galaxies continue to produce stars (they are taking dimensions from chaos and are on the edge of chaos) whilst other parts of space can only add a limited number of dimensions at a time through the touching places on matter, probably between Dq and the three dimensions which are curled up in each particle. The conservation of mass and energy would only be true in parts of space that are not creating stars.... where stars are being made many particles will be added and some particles may also be lost (because of contact with anti-matter)

Where particles are not being made the matter is not in contact with Cahotage unless the touching places, which we will call **enabler**, are activated by the activator. The activator is also called superstring and has one dimension in 3d and one in Dq. When the activator touches the enabler, dimensions are permitted through from Cahotage.

This new theory is already difficult to understand so we will go through it particle by particle.

First particle has three normal dimensions of space (3d) and Dq and is touching Cahotage all round. It also has a line of activator (this is superstring with one dimension in normal space and one or more in Dq) and an enabler touching place.
The line of activator is attracted to the enabler and would therefore create a loop back on itself if allowed to .

However, it also attracts the dimensions necessary from cahotage to capture a new particle. This may be a particle already in the Cahotage or it may be just the dimensions ..... it does not matter in the theory as far as I can tell but will later put this right if necessary.

The new particle captured by the first one has its enabler activated by the enabling superstring (activator) and work is permitted. This consists of capturing "space and time" dimensions. The normal spatial dimensions are lined up in three dimensions by the template of three dimensions inherent in the particles but the "time" dimensions (note my scepticism with regard to the term time here) are incorporated into Dq.
In other words the particles move apart.

It could have been that the first particle met an anti-particle in the infinity of cahotage. If it had done so the superstring of the first particle would have been activator and the superstring of the second would have been enabler, the particle of the first would have had an enabler and the particle of the second would have an activator. There would have been an enormous amount of dimensions let into the universe annihilating both particles and thus ending the short life of that universe.

So we will ignore that scenario and assume that anti-matter will not be incorporated into the universe since every time it is attracted there is annihilation of itself and of the attracting particle.

The two particles attract more particles and so on.

The particles do not always attract exactly similar particles ..... some will have multiple cahotage points and some will have multiple superstrings. They will attract each other more strongly if they have more superstring and more Cahotage enabling points.

All true normal particles would have enabler touching points and would thus have mass and all normal particles would have superstring and therefore have gravity.

Two possibilities are now permitted with the theory :
Either the forces would be completely separate : Lines of gravity superstring would interact with Gravity enablers (Cahotage points), lines of electromagnetic superstring would interact only with electromagnetic enablers, weak interaction would only work with weak interaction and strong interaction with strong interaction.

The second possibility is that superstring activators are identical for all forces and the touching point enablers are identical but they are differently positioned on the particle. Gravity would have centrally symmetrically positioned enablers and several activators. The lines of gravity would emanate out like bristle brushes and would repel each other sideways. This is described in greater detail in the section on gravity.

With magnetic particles the enabler cahotage contacting place would be asymmetrically placed. The magnetic superstring would emanate out from one end like a bristle brush and be attracted to the other end where the enabler points would be lined up. The superstring magnetic lines would be kept apart because they repelled each other sideways.

If a magnet was oscillated the superstring lines would be compressed together and would meet each other to be repelled by the repelling force of "like" superstring. This would thus be propagated as a wave. When the wave met activating points on particles it would permit more work to be done by the incorporation of more dimensions.

(In our present language of science this would be interpreted as a electromagnetic energy hitting matter and heating it etc. The dimensions of energy were ML^2T^{-2} yet we knew that a photon was massless and did not have these dimensions. In our old terminology light should not have been thought of as energy but as a force. Even then the mass referred to is the mass of the body that is affected by the force and not the mass of the force itself)

There is more about magnets, gravity, quarks, gluons and Higg's bosons later in the book.

When the universe reached a critical size, part of it could well have exploded in an exponential way leaving the background radiation of $3°K$. This is not essential for the theory but could explain why the oldest stars seem to be older than the big bang. The pattern of the galaxies is discussed later.

Chapter 12
Fridge Magnets and Superstring
Unlocking the Secrets of the Universe

In this theory electromagnetism is possible because of asymmetrical positioning of the superstring and cahotage touching point .... the activator and the enabler.

Thus the superstring is able to curve back on itself and touch its own enabler. The permanent magnets are collected into little domains of magnetism and it would be possible to construct a magnet that acted like the traditional bar magnet. The repelling action would be a sideways force due to the deformation of superstring by whatever accelerating agent pushed the magnets together. The sideways repelling force of superstring would be the factor that kept the superstring lines of force apart.

The attracting action would be due to
a) The admission of dimensions from cahotage when the activator touched the enabler
a) Once accelerated the direction of the movement would be due to the fact that a superstring was attached to the body and due to the bending of space by the superstring.

But one could also develop a very flat magnet in which the North and South poles were not an obvious feature. The lines of force would emanate from the flat plane of the magnet at right angles to the surface. If you had two such magnets they may attract on both faces, at each end and oppose only at the long edges. That opposition at the long edges would easily be overcome if there was even the slightest overlap of the magnets. Such a magnet could not be created and retain its magnetism if the old version of like poles repel, unlike poles attract were the case. So we have an easy prediction from the theory ..... based on the idea that superstring attracts along its length and repels sideways.

How was I to make such a magnet ? Does such a theoretical magnet exist.
The answer is yes: it is a fridge magnet and it meets all the predictions. It can even be magnetic on one side but not on the other.

This is all easily explained by superstring but not at all understandable by simple bar magnet physics.

CHAPTER 13

What is gravity ?

Goddard's Theory

In our model gravity is superstring emanating like a bristle brush from the object with mass. The bristles or strings come out from every particle and extend for a perfectly round body affected by no other forces into an expanding sphere of strings extending out from the object. The superstring is one dimensional in the normal three dimensions but extends into the Dq along its length.

The extension into Dq permits the 3 dimensions of space to bend along the length of the string. The bending decreases as the string thins out. This will be inversely proportional to the square of the distance close to an object but will not continue to be so if it is very far distant from the object because there will be space between the strings. The strings repel each other at right angles to the single spatial dimension thus emanating staying apart from each other. The repulsion is resistance to deformation rather than an active attraction of "energy".

When the superstring, or activator, from object A meets an object B it will act on it as a multiple of the number of strings from A passing through the object B by the "mass" of the object where mass is proportional to the number of Cahotage enabler touching points. The action of B on A will be the same ... a multiple of the number of strings from B passing through the Cahotage touching points of A (i.e "mass'). The string meeting the Cahotage enabler touching points activates them and permits Length and time dimensions to pass through (acceleration). When the objects move more length dimension is permitted through from Cahotage. The bending of space and the attachment of the superstring ensures that the accelerated object will move towards the other object.

Due to the curvature of space caused by the superstring, even massless photons will bend in towards a massive object ...... they are not affected by gravity as such but are taking the shortest route in curved space. This part of the theory is identical to General Relativity.

The string may have duality properties of string and wave. It may be straightforward to pass a wave down or across string but due to the immense length it may be very difficult to turn it through the dimensions.

The other "forces" may also have duality and be both wave and string but due to their relatively shorter distance of action they may turn more easily through the dimensions and multiple properties could thus be observable.

Rules

a) All activated enablers have a rate of admission of dimensions L^2T^{-2} which has an identical quantum for one enabler and one superstring (activator). (This is equal to or proportional to Planck's constant, h for electromagnetic enablers and may indeed also be so for gravity enablers).

b) The mass of a particle is dependent on the number of "visible" enablers on the particle. Visible means present in our three spatial dimensions rather than turned into Dq.

c) Because a body is made of particles, the mass of a body will also be dependent on the number of enablers

In addition for Gravity

1) The superstring emanates from a perfectly spherical mass into an expanding sphere of string with the number of strings per area closely approximating to the inverse square law i.e. the number of activators is inversely proportional to the square of the distance d from the source to the place at which the intensity is being measured.

2) Classically the work done by gravity is considered to be proportional to mass by acceleration by length.

But the acceleration due to gravity is also inversely proportional to the square of distance d

In Goddard's theory

The work done by superstring is proportional to the dimensions $L^2 T^{-2}$ let through from Cahotage. This in turn is proportional to number of activated enablers which is proportional to the number of enablers by the number of superstring activators. Since the number of enablers is proportional to mass and since the number of superstring activators per given area is inversely proportional to d^2 the relationship is nearly identical to that in the classical theory.

The main difference is that the superstring:

a) Only approximates the inverse square law becoming less accurate the smaller the area chosen and the further away from the body concerned.

b) Bends space/ time according to Einstein's predictions close to the body but space is flat between the strings. Because of the bending of space, the dimensions let through for acceleration times length would have to be slightly greater than one would expect for the length and time observed from a purely three dimensional view point. These increased dimensions result in time dilation and are discussed later. They are already mathematically described by Einstein. Once again the only difference from Einstein is that gravity is bending space rather than gravity being due to bent space.

Newton
Now I have caught this Goddard guy out. Anyway, who is he to tell me all this? He's not even a physicist.

Socrates
Forget the abuse, what is your objection to this theory ?

Newton
There are plenty of string theories of gravity and there were in my day .... we just

called them lines of force. However you play around with the shape of the spatial dimensions you cannot make an inverse square law.

Socrates
Why ever not ? The strings are emanating like a bristle brush and would almost exactly obey the inverse square law close into the object

Newton
But they would not emanate like a bristle brush. They would all be drawn to the nearest large object. I accept that the strings from the sun could emanate in a spheroid due to the effect of our galaxy and other galaxies but even then it would not be exactly right. In fact the galaxies all seem to be approximately on one plane so what you would have is an irregular ellipse.

As for the Earth .... nearly all its superstring would point towards the Sun. We know that this is not the case because of the inverse square law so Goddard's theory must be wrong.

Socrates
But what if the superstring repelled itself in the horizontal plane. Repulsion due to forces is already observed.

Newton
They were not referring to Gravity. They were referring to all the other forces.

Socrates
That's strange is it not ? I'll stick with the idea that the strings repel themselves as an "antigravity" effect. The idea is that the superstring is the agent that warps our three dimensions in the extra dimension D_q. It resists bending in the sideways plane and this acts as repulsion when two lines are pushed together. The same action was seen for the lines of force or superstring of magnets : it is a passive response resisting deformation and storing the ability to do work if it is deformed.

Newton (sarcastically)
Why not call the effect superspring

Socrates (taking him seriously)
Hey, you're getting into the swing of it

Newton
But there are still problems. Why can't we feel the effect of superspring as a sideways force on us.

Socrates
The net result is very close to zero in a sideways plane unless you are measuring at the quantum level between the strings.

Newton

But that would bring a hidden variable into quantum theory.

Socrates
There are almost certainly multiple identifiable hidden variables resulting in the uncertainty of quantum theory. Its not really a theory, it is an extremely sophisticated adding machine.

Newton
But there is always my Universal Gravitational Constant G

Socrates
If you tried to measure the constant of Gravity G you would get exceedingly close to the constant but the results would differ very slightly on each occasion due to three reasons.
a) The strings of the objects you were using to attract one another would not be quite the same number each time.
b) The strings from the earth would not be quite the same number each time
c) The sideways force would not absolutely cancel out.

Newton
But they have made very highly precise measurements of G

Socrates
And they all differ slightly.

Newton
So what is the true constant.

Socrates
The true quantum of gravity is one superstring and we should be able to measure its force by performing gravitational experiments in space or by looking at the differences between the measurements of G. I'm afraid that such measurements will have to be done by others .....I'm off for a drink of hemlock.

Newton
Hey I've found another objection
If there is space between the strings space could be very flat in those parts but Einstein predicted that space would be very bent. Einstein has been proved right by the curvature of light from the stars round the Sun, so Goddard must be wrong.

Socrates
Good try.
Space is curved around masses and the curvature is very close to that predicted by Einstein but over the vast entirety of Space it does seem to be very flat.

Newton
I do not believe it. Einstein can't be wrong. He verified all my laws as special cases
What is your reference source for that .

Socrates
Andrei Linde. Sci Am. Nov 94, page 32

Conclusion

What have we learned from this discourse

In the theory of this book

1) Gravity is superstring or activator
2) It very closely approximates to the inverse square law due to the distribution of the string emanating from the body
3) The string is distributed in an even manner around a spherical object that is unaffected by other "forces" due to resistance to deformation sideways. This can be considered as a "superspring" effect at a perpendicular to the length of the string and is probably due to the extra dimension D_q.
4) The constant G is very hard to measure accurately because of the quantum of gravity ... one superstring.
5) Space is very flat
6) Just as Newton's laws (when they were right) were a special case in Einstein's theory, Einstein's laws are a special case in Goddard's grand unifying theorem.
7) Gravity acts on an object by activating the quantum contact points (enablers) and permitting through the dimensions L^2T^{-2}. This is acceleration over a length and is usually known as energy but we will call it "work done".
8) The dimensions let through by an activator are identical so any "force" or superstring activator will result in acceleration. Acceleration of a body by any activator (force) could have the same quantitative effect as gravity but its qualitative effect could be different : for example, the acceleration of your car does not obey the inverse square law and can be varied at whim.

CHAPTER 14

Time and Motion Studies

Time is much more complex than just a single dimension.

We will have to distinguish here between

Objective time (spatial time), that which is measured by a watch,
Subjective time, that which is experienced by a conscious observer
The arrow of time, the unidirectional build up of Entropy in the Universe
Available time, that which is immediately available to be used by the object
Past time, the memory of experiences
Future time, the prediction of future available time

All of these are different for each observer if they are not exactly moving together and thinking together (which is probably impossible).
Time and motion are intricately linked. The motion may be acceleration, velocity, heat or sound .... the latter two being special cases of motion.

So off we go:

In the classical theories a force would be required to move an object . Work would have been done by release of energy. When the force stopped pushing, the object mysteriously carried on moving due to momentum, unless stopped by another force. But this did not explain momentum which was just the name given to an object that is mysteriously moving at a fixed velocity. No explanation was given to explain:
a) What is a force?
b) what is energy ?
c) Why should the object move when the force stops pushing it ?

The dimensions of the above are described in terms of mass, length and time. But the mass is the mass of the object being accelerated not the mass of the energy required for acceleration.

The Goddard theory is completely different but still agrees with the mathematical results.

When considering magnetism, either the required dimensions are stored in deformation of the superstring (which repels sideways) when work has been done to bring the magnets together or the superstring must have contacted the touching places on particles. In our simpler jargon the Activator must have opened the Enabler . (Do not forget that the superstring of antimatter would be Enabler and touching place Activator.)
The latter situation is identical to gravity.

When the touching place is activated dimensions are permitted through ready to be used as work done. The greater the acceleration or the velocity the greater is the amount or quantity of dimensions let through. There is always a small net increase of dimensions over the amount that is then used. We have gained available time

But what has happened to the arrow of time ?
That which is used to move the particle over the corresponding space in our three normal spatial dimensions passes into Dq as something very akin to Entropy, which we shall call Extropy. This corresponds to the arrow of time and happens at the particle level just as it does at the macroscopic level. Apparent reversibility or not of the motion is irrelevant .... because it never is reversible in this universe! When a particle moves it will have used dimensions which are then added to the Extropy of the Universe.

The available time and space that are not used are still available to be used and because greater acceleration or velocity capture greater quantity of dimensions this is manifest as time dilation and increased mass. In other words because the time and space are captured from Cahotage the faster the object moves the more time and space are available. The moving object has made time available ! This occurs in this theory exactly as it does in Einstein's theories of relativity : the only difference is that it is the gravity that bends space not the bent space that is gravity and it is the movement of the particle with its attendant three dimensions, Dq, superstring and enabler that bend space when it is proceeding at a uniform velocity. The size of the opening to Cahotage must be squared when the object is accelerating compared with velocity since the dimensions let through are LT^{-1} for velocity and L^2T^{-2} for acceleration and are markedly increased for rate of change of acceleration or for complex volume acceleration in multiple directions as seen in an explosion. It follows from this logic that the massive expansion of dimensions from an explosive emanate from a large rent into Cahotage (or chaos). The larger the explosive force the larger the rent. The potential ability to release an explosion would then be due not to massive stored energy in the chemicals but due to available enabler and activator points being brought together. It would also follow that the energy released when an atomic bomb is exploded is not due to the energy stored in a few tiny atoms being released but due to more efficient activation of enablers by activators : in other words an even larger rent in our three dimensions and Dq and a larger contact with the maelstrom of Cahotage or chaos.

The increase in Extropy that occurs is, therefore, greater if the object has moved than if it had stayed still and therefore not only has more time been made available but more *arrow of time* has been used which seems like a paradox but is not. Also the increase in Extropy is greater the more acceleration, rate of change of acceleration or complex disordered motion (such as heat) is used.

The subjective sense of time is very complex . It is related to both the arrow of time and objective time dimensions and is occurring at the quantum level in the brain due to the movement of particles. It is affected by all manner of activities in addition to the objective measurements but appears to be aware of time being made available and time passing as the arrow of time. Thus when experiencing something that is very exciting there are complex appreciations of time. Whilst doing the exciting thing time seems to stretch out in front of us and it seems like it will never end. But when it has ended we look at our watches and say ...... gosh, hasn't time flown by ! The stretching out in front of us would seem to relate to time dilation but when we are back in our usual time frame we notice the passage of the arrow of time. The same thing can happen when reading an exciting book because the imagination is all that is required to move the electrons around inside the brain and capture the time dilation and arrow of time aspects at the quantum level.

But why didn't the watch of the observer experience the same subjective effects. Although on a roller coaster it experiences the objective change of time available, the subjective time sense is related to the excitement felt and although that is partly due to external influences it is much more an internal experience. Reading good books never loses its thrill but going on a roller coaster does.

An observer on the Earth would be affected mostly by the Earth's own acceleration and movement. For him time would have been made available at a standard rate and would not have dilated. He will see the other observer moving at near light speed apparently taking an enormous length of time : indeed to the Earthbound observer the distances that the other observer moved could only have been traversed in a very long time. When the other observer returns to Earth his colleague on Earth will have aged more than him. The subjective impression of the observer who has returned will have been that initially the trip stretched out in front of him with enormous amounts of time available but now that he has returned, he remembers it as time past in which a lot happened but it has obviously all passed in a flash. He will look at his colleague and say," My goodness, haven't you aged, how time flies !"

Now all that is written above will fit in with the calculations of Einstein's two relativity theories. But for the first time we have managed to tie in the subjective sense of time and feel happy with a theory that otherwise rebelled against our senses.

INTERLUDE
ENERGY ?

Newton

I'm still very confused. Goddard seems to think that things can move without energy. We know that this is not the case. Energy is either being used as kinetic energy, which can be ordered or disordered, or it is stored as potential energy.

Socrates

That is indeed the standard description of what happens. However a moment's thought will show you that there is dangerous confusion about the concept of energy.
People do not just think of energy in the terms described. They use it in general conversation (I've got no energy this morning). They also call the electromagnetic radiation from the sun "Energy" although we know that the work done in, for example, the heating of my skin by the sunlight occurs at my skin surface and not in space. The "energy" of sunlight is not released until it hits an object.

Newton

OK, so I would call sunlight a force which can permit work to be done.

Socrates
You could do so but I would prefer to call it electromagnetic radiation or, even better, waves transmitted by the deformation of magnetic lines of force or superstring.

Newton
But you have not told me what is wrong with the terms potential energy and kinetic energy

Socrates
Energy is purely a concept invented to describe what happens when a body is accelerated or decelerated and various other combinations of movement occur, some of which may be disordered such as heat. It has been a useful concept but was wrongly defined initially.

Newton
In what way

Socrates
When you deform an object such as a spring and hold it in place by a catch, you have indeed stored the ability to do work and this could be called "potential energy". Unfortunately the units of energy are peculiarly wrong since a new object that you now place on the deformed object could be said to possess the potential energy depending on the mass of the new object. When the spring is released the new object is accelerated and is said to possess kinetic energy.

Newton
Yes I agree

Socrates
But the body you placed on that deformed object did not possess that energy as potential energy at any time. The second object and the first could not possibly have both possessed the potential energy. The deformed object possessed the energy and the second object possessed its own mass. Releasing the spring allowed the spring to regain its original shape and in doing so conferred acceleration to the second object that was resting on it.

Newton
Let us try it with two bodies tied together by elastic

Socrates
Pull them apart and indeed they possess the ability to accelerate back together and, just like the deformed spring they did possess the ability or potential to do work and it was then put into action. In this situation energy can be replaced in the sentence by the potential to do work

Newton
I still like the term potential energy. My best example is lifting a body against Gravity. It then possesses potential energy does it not ?

Socrates
In classical teaching one would say that it does.
However a body moved away from another body is in a "lower energy state" according to some ways of looking at the situation. The further away from each other two bodies are, the easier it becomes to separate them because the acceleration due to gravity is decreased according to the inverse square law. This is particularly obvious when we look at the energy states of orbitals of atoms. The closer in to the atom an electron is , the harder it is to remove it and the further out it is the easier it is to remove it.

Newton
You cannot get away from the point that if I climb an apple tree ....

Socrates (Interrupting)
Or like Galileo before you, the leaning Tower of Pisa

Newton
Yes, if I climb up with an object in my hand it has stored potential energy. When I let it go it will fall to the ground.

Socrates
No, it has not stored energy. When you let go the force of gravity will act on the body.According to Goddard's theory the superstring will permit the dimensions required for acceleration through from Cahotage by an activator/ enabler reaction

and the body will be accelerated.

Newton
I do not see why it is not exactly like the elastic.

Socrates
Because the higher you go the weaker the acceleration due to gravity becomes
This is not obvious on Earth because you cannot get far enough away by climbing a tower. It is obvious if you send a rocket into space.
The further apart you take two objects joined by elastic the greater the deformation and the greater the stored ability to do work until you exceed the elasticity and the elastic breaks

Newton
I would prefer to stick to my concept of potential energy. It much easier to understand

Socrates
But it is completely wrong. Let me try again.
if an object is lying on the ground, does it possess potential energy ?

Newton
No

Socrates
Would it be possible to give this object potential energy without moving it or touching it in any way ?

Newton
No.

Socrates
If I now dig a hole right next to it, I could push the object over the edge and it will accelerate due to Earth's gravitational field

Newton
You must have given it potential energy by digging the hole.

Socrates
No, that is fundamentally wrong. At no time did I give that object any energy at all. The concept of energy is so abused that it should not be used.
Let me resort to the mathematics to sort this out.

Einstein showed that $E = mc^2$ and Newtonian theory had already shown that Kinetic Energy = $\frac{1}{2} mv^2$.

So the dimensions of energy are $M.L^2T^{-2}$.

But when you move an object of mass m up a tall tower a distance of h it is said to

possess the potential energy mh or mass by the height it is moved through.
These are the dimensions ML. They are not the dimensions of energy.
It does not possess more energy.

Newton

It has the potential that if I push it over the edge it will accelerate to the ground

Socrates

That is true but it does not possess that acceleration until the acceleration due to gravity acts on it.

Newton

Yes but everything that goes up, must come down.

Socrates

That seemed to be true in your day but it is clearly not the case now.
Voyager 3 is on its way out of the Solar system and many rockets have left the Earth never to return.
The further away an object is from the Earth the less the acceleration due to gravity is due to the inverse square law. Thus if you lift a body up into space it has both a potential to come back to Earth and a potential to fly away from the Earth. It possesses no less Energy and no more Energy than if it were sitting on the Earth but its potential future actions may have changed. There are other potential future actions we could imagine : it might explode, it has a very small chance (which increases the smaller the object becomes) that it may move to the other side of the universe as permitted by Quantum theory (In Goddard's theory :due to folding of space or movement through Dq). Alternatively, it may stay in geodesic orbit, especially if I send a shuttle up every now and then to give it a nudge back into orbit.
It does not possess a mythical potential energy.

In Goddard's Theory Movement occurs in three ways

1) By the activator / enabler reaction ... this permits space and time dimensions to enter the Universe from Cahotage. Slightly more enter than are required for acceleration or velocity thus causing time dilation . Multiple dimensions of time may be used for complex acceleration, rate of change of acceleration etc.
2) By the bending of superstring storing the ability to do work or transmitting a wave (the activator / enabler action is required to set this off).
3) Apparent movement ... the bending of the path of "massless" particles such as the photon when they pass a massive body is due to the bending of space in Dq. The path of the light ray is not really bending .... it is taking the shortest route in multi-dimensional space .

CHAPTER 15
Quantum mechanics

In a way this entire book is about Quantum Mechanics since it is an all pervasive subject. However, it would seem worthwhile to draw together some of the points specifically in a short separate section.

Some of the different factors of quantum mechanics were put forward in the first section of the book and could be reiterated here.
a) The quantum of energy is frequency x Planck's constant
b) The principle of uncertainty involves local and distant components

In this theory a single superstring can activate a single touching point. One unit of activator can activate one unit of enabler. This then permits dimensions through from Cahotage.

But for electromagnetic radiation the number of strings hitting one point is dependent on the frequency. So the quantum of electromagnetic radiation would be frequency multiplied by the constant for the touching point multiplied by an unknown factor. The formula is identical to that of (a) above. The Planck's constant is therefore presumably the constant for the touching point but there may be a further additional factor.

Hidden variables would seem to account for a great deal of the uncertainty
1) Local causes of probability include chaos in the means of production: the production device must consist of multiple particles and is therefore a chaotic system. It also obeys the relationship of superstrings multiplied by enablers because work must be done to produce the electromagnetic wave.
2) The position of the enabler on the screen or on the detector. You could not see the "photon" unless you had an enabler on the detector so this can explain the peculiar differences when you can either demonstrate the wave or the particle.
3) Hidden mathematical information inside the particle (eg. electron) invisible due to the voxelation (or manifoldation) of the spatial and time dimensions. There is an experimentally proven quantum of measurement of length and of time beneath which it is not possible to resolve a difference between measurements.
4) Distant effects due to Dq : the extra spatial dimension(s). These would explain quantum tunnelling and probability of distant position, Einstein , Podalsky , Rosen entangled pairs etc and some aspects of duality such as the flavours of quarks.
5) The remote possibility that consciousness affects it due to quantum tunnelling or direct bending of space.

Rules

All activated enablers have a rate of admission of dimensions L^2T^{-2} which has an identical quantum for one enabler and one superstring (activator). This is either equal to or proportional to Planck's constant, h.

Emission of electromagnetic waves would be dependent on the enabler/activator reaction at the site of production of the radiation. Considering the production of X-rays, *Bremsstrahlung* or "braking" radiation, due to interaction of a moving cathode ray electron with the target, will serve as an example. Two types of interaction between superstring and enablers are envisaged.

In the first the moving electron's superstring reacts with the superstring of the electrons of the target. This sideways deformation of the superstring (deforming the strings LDq) causes repulsion and transfers some of the velocity of the electron to the atom as disordered movement or heat. The moving electron moves off at a different angle and with a lower velocity.
(In classical language the electron impinges on the electron cloud of the atom and is repulsed).

In the second interaction the electron penetrates past the atoms outer superstring and meets the enablers and activators of the nucleus. When it reacts with the latter the enabler/activator interaction permits dimensions to be used in the opposite direction to that of the speeding electron. These will be only admitted at a quantum rate but will decelerate the electron and in doing so permit the production of a wave across the magnetic lines of force emanating from the nucleus. The greater the number of activated / enablers that decelerate that moving electron the greater the frequency of the wave. (In classical language the greater the loss of energy from the moving electron the higher the energy of the photon of X-ray produced). The spectrum for a number of moving electrons would , however, be "continuous" because the loss of velocity (the deceleration) will depend on how many strings and enablers managed to meet each other. There will clearly be a peak but an electron that escaped being decelerated completely will move to another atom where it will lose velocity and the frequency produced will be lower. If there is only one moving electron it could have a peak production of X-rays due to all its possible enablers and activators contacting the activators and enablers on the nucleus and being completely decelerated to a halt (head-on collision) thus producing a packet of waves with the highest possible frequency for that initial velocity. This would be the same as the frequency being related to kVp.
Note that more electrons of the same velocity would **not** increase the maximum frequency and that wavelength is inversely proportional to frequency.
More moving electrons would increase the overall intensity (or quantity) of the X-ray production and produce a spectrum but would not alter the quality (kVp).

For absorption of electromagnetic radiation let us again consider X-rays. The quanta of dimensions let through for work to be done would depend on the number of superstrings touching the enabler of the film per second and the constant for an activated enabler (proportional to Planck's constant).

Since in this theory electromagnetic radiation is a wave across superstring the number of strings per second hitting an enabler will be proportional to Frequency. Therefore the quanta of dimensions for work done is proportional to Frequency multiplied by Planck's constant.
The same relationship would apply for any electromagnetic waves.

In this theory the electromagnetic waves are produced in quantum packets of waves but the observed particle nature lies in the detector not in the packets of waves.

CHAPTER 16
Weak interaction, strong interaction and the Higg's boson

The theory espoused in the second half of this book has not, as yet, said anything about the other two forces of classical theory: the weak and strong interaction. But it will predict features.

They will have the dimensions of one length (maybe straight or curling back on itself) and Dq. They will therefore be a form of superstring and will be attracted via their cross-section (zero dimension) to a cahotage touching point on a particle and will release dimensions because of the attractor enabler reaction.

As with the other "forces" they will repel at right angles to the length of the superstring when the shape is deformed. The weak interaction and strong interaction could be described as a wave (similar to electro-magnetic spectrum) but, just as with the other two enablers, the observed particle nature is due to the activator. The wave/particle of both the weak and strong interaction would have no mass in the same way as the photon or gravity have no mass.

Because the superstring emanates from a particle the "particle" that theoretically represented the force would not exist separately from the particle that emanated it.

As far as my reading permits in this very difficult area this description would fit exactly the theoretical particles that carry the weak and strong interaction such as the elusive Higg's Boson that only the most advanced theoretical physicists like to even think about.

Naturally they would be subject to quantum measurement just as everything else is. Again the quanta would be one superstring and one cahotage touching point i.e. one activator and one enabler.

Good luck in sorting out this area !

Newton
Ah, ha. I've got him now. He said that the Higg's Boson would be like electro-magnetic radiation and have no separate existence from the particle it emanated from. Well I know for a fact that light is made up of photons of a fixed energy depending on their frequency and Planck's Constant.

Socrates
You are forgetting what you've been reading. Light is a wave of electro-magnetism. It is a distorting wave across and along the superstring of magnetism (the lines of force). When it meets a particle it activates the Cahotage touching point (enabler) and permits a certain amount of dimensions through permitting work to be done. The amount let through will depend on the number of superstrings touching the enabler and the constant of the enabler. The quantum of work done ("energy") is therefore a product of frequency by the enabler constant. This relationship was described by Planck but just appeared in Dr Goddard's theory by logical analysis. Nearly all the important measurements have been made and it is

only interpretation that is so difficult.

Newton
I would prefer to agree with Bruce Gregory. He's got loads of degrees in the subject and he thinks that physics is only a concept of language and that physicists do not discover the physical world, they just invent it (*Inventing Reality, by Gregory, an excellent book Ed.*).

Socrates
Oh, dear. He is so close to the truth but yet so far. If we accept that the only thing I can know is that I exist (*Cogito ergo sum*) and the only thing that we next need to believe is that there is an outside reality then his thesis must be the wrong way of handling the situation. If there is an outside reality we should try to find the best model to fit it and it seems that Dr Goddard's model is the one that predicts the greatest number of facts correctly.

That does not make it right. There may be a better model and there are enormous areas that he has left out. What is this cahotage or chaos that surround the universe in his theory? Are there other universes in it? What about chemical reactions, different atoms and molecules. Its not complete but it seems to be a better model.

Newton
But its not written out in mathematics

Socrates
No, but the logic is sound

Newton
But its just a load of words. Without measurement there is no science

Socrates
If I cannot logically convince you on that one, then we will just have to disagree

Newton
Heh, wait a minute, you're giving up on me.

Socrates
The examples of scientific advance that have occurred by chance and not by design, by luck rather than judgement and by trial and error rather than by measurement are too numerous to mention. Once the discovery has been made measurement is used to confirm it. Blind measurement alone has rarley led science forward. The same applies to elegant logical theory : once having been put forward it should be tested by measurement, verified or refuted. If the latter, it should be rejected or modified.

Newton
Pass me your hemlock

CHAPTER 17

Mathematics, The Mandelbrot Set and the Pattern of the Universe

Caveat

I would like the mathematicians to treat this section as science fiction : suspend your disbelief and see if there is any internal logic once the original rules have been set. I apologise to teachers and professors of mathematics everywhere : I'm really on your side . Mathematics is the last bastion of syllogistic logic, even the logicians have moved away from it. If you cannot read this section, do not do so. It does not matter with regard to the rest of the theory presented although it ties in extremely well.

Mathematics is a way of presenting logical arguement. It is a symbolic representation of logic. There are four main problems with mathematics:

1) The way that mathematics is written is fine for mathematicians but is a bar to understanding the logic behind it a) for non-mathematicians and b) for mathematicians themselves. If all the problems had to be written out as logic the answers would be understandable to all logical people. There is no point, for example, including 17 dimensions in an equation "in order to make it work" unless you have some idea of what you mean by the 17 dimensions.

2) The names given to things alter the way we think about them. A good example here is real and imaginary numbers. The imaginary numbers are just as real as the real numbers and are used by mathematicians and engineers all the time, but try explaining them to a non-mathematician and they do not believe they exist .... because they are said to be imaginary . It would be far better to call them conceptual numbers.

3) Chaos theory is fascinating and upsetting. One professor told me ..... "It tells us more about what we cannot know than about what we can know ". He felt that its star was on the wane.
Mathematicians have an aversion to chaos theory because there are a large number of different systems to describe turbulence, uncertainty etc. without need to resort to chaos theory . But contemplate number 1 above..... the logic may be right and yet the theory does not work

4) Most logic ,and therefore mathematics, does not perfectly describe the real world. The logic seems to be too good for reality. This is hard to accept but is probably true due to all the same causes of uncertainty as are seen in Quantum theory and chaos theory.

My worries about mathematics conveyed to the reader we shall continue ....

Logic and the real world

Logic can transcend the physical world ... so mathematics is not bound by physical limitations. Mathematics is a language of logic and logic is not constrained by the shape and dimensions of the universe. This can be both delightful and frustrating. Chaos theory has stated that once there are more than three variables the system is non-linear and that however well one can imagine the situation being perfectly repeated ..it never is. Thus the imagination of a mathematical reality is never attained.

We will take an example. We can mathematically describe a perfect circle, a perfect traiangle and a perfect square and many more shapes. We can mathematically describe a perfectly repeating fractal or the perfectly variable infinite fractal set known as the Mandelbrot's set.\

But are they achieved in reality ?
At certain levels in the system they may appear to do so but none of the examples given above are achieved at all levels in the system.

Mandelbrot Set

$$Z \rightarrow Z^2 + C$$

Where Z is any number starting at zero and C is a complex number.

The algorithm is very simple but the result is astonishingly complex and amazingly beautiful.

If the Mandelbrot set is a unique set of infinitely repeating but infinitely variable two dimensional patterns then if I have a two dimensional pattern in front of me, it must be included in the set somewhere.
This may or may not be useful to me because I may have to search for billions of years to find the appropriate set. Even if the pattern would be high up in the set I may not recognise it as being there because all the superstring would have to be included and this may only be partially visible if demonstrated by stardust, planets, iron filings or whatever. In which case the remaining pattern may be well down the set. However, whether we view the stars in their galaxies or the turbulence in water we will find that universality has crept in and the Mandelbrot set is lurking. This is very upsetting for the mathematicians. They can, for example, describe a perfect pendulum or even logically demonstrate mathematics of non-linear systems other than the Mandelbrot set or chaos. In reality they never work perfectly because at the edges chaos is lurking (friction at the pivot, air resistance, quanta of gravity, the voxelation of the universe) . Even still, it can be better to use the non-chaos mathematics in many situations and add what the astronauts called "a fudge factor" because the results are usable.

The perfect geometric shape or repeating fractal cannot be drawn because at the edge of the shape or fractal there is always irregularity when magnified sufficiently. This in its turn is a repeating but variable fractal like the coast of Britain.

At the quantum level objects may appear to be perfectly identical. But this is more related to voxelation (or manifoldation to coin a new word) of the universe than to the object , such as an electron, itself. A voxel is a three dimensional equivalent to a pixel and manifoldation is the multiple dimensional equivalent. The spatial dimensions and spatial-time dimension have a quanta of measurement which is the smallest discernable measurement. Below that measurement objects may be present but will not be fully shown. If they have sufficient mass, charge etc. to be shown they will "fill the pixel" when we measure them even though they are only taking up part of the volume. There may also be a lowest and highest amount of dimensions available from Cahotage to allow work to be done. This also may depend on the equivalent of the range of values displayed. (See also glossary on CT, partial volume effect, voxel and pixel.) A small, less than quantum sized manifold will appear to be the same size as another because of the smallest available spatial /time dimension. Objects that are larger than a single quantum measurement manifold would also be affected by the partial volume effect.

An object that is 3.9 pixels in size would fill 4 pixels in exactly the same way as one that was 3.854 or even 3.6 and the same would hold true for complex manifoldation of the universe at the quantum level.

Thus all elementary particles could appear perfect and of identical size even though they could vary slightly below the "pixel" of dimensions.

The identical nature of the particles is in contradistinction to the idea that the Mandelbrot set could be the pattern of the universe ... but the manifoldation of the spatial and time dimensions would prevent the very small differences from being resolved and the mathematics of the set could continue unabated. Particles could be smaller than the quanta of dimension and also too small to be initially detected. The mathematical description of the inside of the "pixel" is not discernable on the surface pattern. If an object is in the quantum manifold it will either show as a pixel or not ... an all or none situation. With a CT of the chest, the blood vessels out at the periphery are 0.5-0.7 mm in diameter but on a CT scan with 1mm by 1mm pixels if the vessel shows at all it will perfectly measure 1mm or 2 mm in diameter. A similar principle is probably acting at the quantum level and making the particles look identical and even though there may be an alteration in their values this is not discernable due to the manifoldation of the universe.

CHAPTER 18
The Fractal Universe, From Galaxy to Quark

The possibility that the universe may be described by using a fractal pattern may have struck many people because of the commonly noted similarity between the structure of the atom and the solar system. The similarity does not stop there. The Milky Way, itself rotating, is surrounded by orbiting spinning satellite galaxies. At the other end of the scale the atom is made up of electrons in orbitals around a nucleus of neutrons and protons. The neutrons and protons are, in turn, made of sub-particles (quarks glued together by gluons.) possessing electrical charge and another form of charge called color and other properties known as flavours. These according to some theorists (including Goddard) are joined together by superstring. At all levels there is an echoing of the solar system model although the pattern is not identical. The observations are clearly similar to the increasing complexities of fractal patterns as displayed on a computer.

Multiple Universes and Fractal galaxies

Andrei Linde in Scientific America November 94 and previously (Linde, The fractal dimension of the Inflationary Universe, Physics Letters B vol 199 No 3 pages 351-357 Dec 24 19879), has used a fractal model to explain inflation after the Big Bang and suggests that multiple universes may be connected by wormholes. In his version of the cosmos the universe appeared out of chaos and the Big Bang was just one of multiple big bangs constantly happening. He suggests that the rules of each big bang could be different and he believes that our big bang may be the result of a random or chaotic evolution in which at some time a universe with the particularly stable nature of our own will appear and persist. In his latest theory Linde states that rather than being an expanding ball of fire the universe is a huge, growing fractal consisting of many inflating balls which in turn produce more balls, ad infinitum. Clearly Linde's theory of the cosmos is approaching exactly the same result for the space-time continuum as Goddard's theory (Goddard P, Fractal Paradigm, New Scientist, letter 17 Dec 1994 pp 50 and 51) but from a different direction and his description is much like the appearance of the first image of the Mandelbrot set.

In his article in Scientific American Andrei Linde suggests that his theory may alter the way in which we look at the world including even our own consciousness. The author would agree that this is the case and from his own entirely separate observations and deductions suggest that the fractal nature of reality extends to all of the observable space-time continuum even now well after the inflation. If the fractal nature of the cosmos was apparent at the early period of the big bang why should it have disappeared and what made it go ? It would be easier to accept that the pattern is still present if less obvious.... if the inflationary period was fractal (if indeed there was an inflationary period) the nature of fractals is for there to be a repeating pattern at every level of magnification. Clearly Linde is describing a self-similar but not self-same fractal and the Mandelbrot set is just such a classification of self-similarity with the next level of complexity resembling but not identical to the level before.

The distribution of galaxies across the cosmos might take on the form of a fractal according to Robert Osserman (Robert Osserman "Poetry of the Universe: Mathematical Exploration of the Cosmos" Anchor (USA), Weidenfeld (UK).). This is interesting evidence that the fractal pattern emerges at the galactic level.

In future copies of this book there will be photographs of Julia sets from the Mandelbrot set and photographs of galaxies. The examples are numerous but all of the galaxies bear resmblance to parts of the Mandelbrot set.

Photographs of galaxies bearing an uncanny resemblance to Julia sets (Outer Space: the Boundless Sky from The Great World Atlas, Reader's Digest, London)

In a photograph of two interacting galaxies known as NGC 5216 and NGC 5218 ,The Guiness Encyclopaedia , Guiness Publishing Ltd, 1990 UK page 82) there is an interesting feature. On ordinary optical photographs the two galaxies appear to be completely different systems but the photograph from a charge coupled device revealed a thin "rope" of stars pulled out of the two galaxies by their gravitational pull on each other.   As described above Julia sets are fractal images that are present within the Mandelbrot set and the islands of the Mandelbrot set have been shown to be joined by a thin thread of computed numbers. It is tempting to suggest that this is equivalent to Superstring. Further evidence of the fractal pattern of the cosmos was evident in photgraphs in Focus April 1995 p22.

Is the Fractal Pattern of the Universe Emerging ?

We may postulate that the fractal pattern of the cosmos is emerging but we must remember that we are dealing with a metaphor or simile. The pattern of the universe may be a Mandelbrot set or multiple overlapping Mandelbrot sets but it would be a dynamically changing fractal or fractals with the pattern emerging in three dimensions rather than two and different levels of magnification visible all at the same time. . The sets would have been contorted by changes in the time and space dimensions in an analogous way to "rubber mapping" on a computer. We would not be able to see an exact pattern from higher up the Mandelbrot set unless we were looking at just one level of magnification ... perhaps we could look at stars of the same age and assess their distribution and show that the pattern is predicted by the Mandelbrot set but perhaps we would do better to look at smaller structures.

It can fairly be argued that we know why the planets orbit the sun and galaxies orbit other galaxies .... it is due to Gravity. But whilst that is a good working description of **what** happens it does not tell us **why** it happens nor does it explain what gravity is. This has, of course , been thoroughly discussed earlier in the book and, in Goddards Theory, gravity is superstring (activator), Cahotage points (enabler) and the bending of space.

The world around us

There are examples of chaos and fractals too numerous to mention in the world around us at both the molecular level and in our everyday surroundings. Examples include turbulence, clouds, the patterns of trees and ferns, the branching pattern of the bronchi and blood vessels in the lungs and the surface of the brain. Many of these examples are discussed and pictured in *Chaos, making a new science,* James Gleick , Heinemann ltd UK, 1987,1988, 1990 . This should be read by anybody interested in Goddard's theory

If a fractal pattern appears at any level it is worth remembering that it is the nature of fractals to go on repeating themselves at lower and lower levels with self-similar but not necessarily identical patterns.

Particle Physics and fundamental particles

It has been the history of particle physics throughout the twentieth century that further and further layers of complexity have been discovered when analysing the "fundamental particle". This is apparent from the atom, philosophically described by the Ancient Greeks but "discovered" at the beginning of this century. This led to the demonstration of the internal structure of the atom consisting of protons, neutrons and electrons and latterly to the further splitting of the protons and neutrons into quarks , of which there are six types. A pattern of hadrons may be built up based on combinations of different quarks and, in an analogous way to that of Mendeleyev's table of chemical elements predicting new elements such as gallium and germanium, the table of quarks can be used to predict the existence of new particles such as the omega particle. This is another example of a self repeating pattern at different levels of the paradigm.   Over 200 elementary particles are now known ! To demonstrate the more esoteric particles requires increasing levels of energy and they are present for shorter and shorter periods of time.

Newton
This strikes me as being evidence by analogy. I do not like to see arguments based on analogy.

Socrates
That is why the theory has been presented earlier. The simplest and most convincing evidence is
1) The prediction of the relationship of Planck's constant to frequency
2) The flatness of space over very large measurements
3) The inability to measure G precisely
4) The action of simple bar magnets and fridge magnets (any one can try that one at home, it does not require sophisticated equipment)

The analogies are minor pieces of evidence.

Newton
But how does Goddard's theory explain all the different particles . Throughout the theory he has referred to single particles with each with a single so-called Cahotage point and a single superstring.

Socrates
Yes that is the smallest complete unit in this theory and is considered to be an elementary particle which would have mass.

All "particles" without mass (photons, gluons, Higg's boson etc.) would be purely superstring and acting in three possible ways :

1) as simple or complex waves
2) by deformation of the superstring
3) by activating the touching points (the activator/enabler reaction)

This latter reaction is the one which gives the superstring its apparent particle nature when it is detected, hence they are often described as being massless particles.

An elementary particle without superstring but with an enabler touching point can be imagined. A swapping of the enabler and activator so the the superstring becomes activator and touching point becomes activator would also be possible and would represent anti-matter.

Newton
But you still have not answered my question.
How can there be so many particles ?

Socrates

How can there be so many planets ? How can there be so many stars ? How can there be so many galaxies ? The answer is that the smallest building blocks of elementary particles will add up to make larger particles and eventually atoms , molecules etc. All of the properties can be worked out in this way but one must remember the sections on Quantum mechanics, Uncertainty, relativity etc. All of these activities occur in this universe based on the simplest building blocks

Newton
So Dr. Goddard is finally a believer in the atom theory rather than infinite divisibility.

Socrates
I don't think he is. I think he believes that the answer to the conundrum is presently unknowable because of the quanta of measurement : there is a smallest measurement of length and a smallest measurement of time that can be made. As he said earlier in the book this results in voxelation or the equivalent in four or more dimensions (manifoldation) and one is unable to detect exactly what is happening below that level although it is sometimes possible to gain general information that

suggests that there is more going on than one thought.

Newton
So it all comes down to uncertainty again.

Socrates
In one way it does. There will be uncertainty in every measurement made because of the quanta of dimension, the superstring, the Cahotage points, the inherent chaotic fractal nature and the folding of space. But now that we have suggested some logical causes of the problem we should not let the uncertainty cloud our thinking. The important aspect is the required balance between chaos and order for all movement and therefore for all life.

Newton
I'm still confused about movement

Socrates
I'm not surprised
Three causes of movement have been identified
1) Movement due to the activator / enabler reaction permitting dimensions in from Cahotage. When this occurs in a single direction due to "aligned" superstring the movement will be acceleration in the direction of the pulling string. The direction is always attractive because the string is attached to the enabler and any movement away would distort space more. When the superstring is curled or disordered the movement can be explosive with dimensions in every direction.
If a body is moving at a steady velocity (i.e. with momentum) the opening to Cahotage stays open and permits through very slightly greater dimensions of space and time than are required .... thus the greater the speed the more the dimensions let through. The dimensions are retained in the universe as a form of Entropy which Dr Goddard has called Extropy . This maintains the arrow of time.
2) Distortion of superstring (which resists distortion because it has the extra Dq dimension) ..... this is the case for electromagnetic waves and when I move a bar magnet close to another bar magnet with so-called opposing poles.
The work done to store that "potential energy" in the distortion of superstring in the case of the bar magnets came from the activator/ enabler reaction in the muscles of your arm. To send electromagnetic radiation the work has been done at the source of the radiation ... for example production of X-rays by *Bremsstrahlung* deceleration of electrons.
3) Apparent movement : this is the case for light waves. They are actually moving due to distortion of superstring as in (2) above . The apparent movement occurs when they swing in towards a massive object because of the multi-dimensional nature of space . The light wave is moving in a straight line but the straight line is through three spatial dimensions and Dq.

Newton
I prefer my laws of movement. They actually add up and are useful for calculation.

Socrates
Except your third law

Newton
Sorry about that one

Socrates
But the other laws are very good and do indeed describe a particular situation. They cannot be extended to high velocities, acceleration, rate of change of acceleration, explosive disordered movement etc. as can Goddard's Theory

Newton
But he has not put the mathematics to it.

Socrates

But he has demonstrated excellent logic and an ability to predict and explain unusual results. This can be very easily translated into physics and mathematics terms by the professional mathematicians. Goddard is not claiming that he has worked it all out, just that his conceptual model is a better way of understanding reality.

Newton
Has he worked out the Quarks yet ? I understand that the quarks vary partly because they are said to rotate in an abstract imaginary mathematical space producing up, down and strange quarks depending on which face of the quark is turned into the imaginary space. How does Goddard's Theory explain that ?

Socrates
Oh dear, the use of the word imaginary when they should have used conceptual. The "imaginary space" they are referring to is known as Dq in this theory and time in the General Theory of Relativity. It would be better to remove the term "imaginary" from the mathematician's vocabulary.

All of the Quarks can be considered as particles with a number of strings and touching points which are being rotated round in the three dimensions and Dq. When a string or touching place is in Dq it does not impinge directly as a "charge".

Newton
I'm finding this all too much.
I liked the Age of Reason

Socrates
But do you presently enjoy the Age of Uncertainty and Relative Values? What we need now is an understanding of the need for balancing order and chaos. We must not ignore the good ideas that arise from chaos but they need sorting out. Balance is required for true progress.

The Summing Up

Some people turn straight to the back page of a thriller to "find out who done it". If you are such a person you may find this summing up hard to understand. If you are not and you have read all the preceding pages you will not find this section difficult. The theory, like all good ideas, is really a synthesis of ideas from everybody else but the author did not realise that when he started to work it out : it just appeared after eleven years of hard work and thought. However, many references to similar ideas do surface throughout the scientific literature. I have not been able to find them succinctly and intelligibly summated in this form before but where I have found echoes of the theory I have included a name. Where the names are included they are not supposed to signify that the way in which I have interpreted the theory was implicit in their work . I am only suggesting that there were hints of the same theory.

The theory is as follows.
The universe evolved out of chaos as the particular universe with the correct factors to permit expansion into the present universe that we live in. This was a fractal nature (Linde).
The first particle attracted other particles from the surrounding chaos of dimensions which we have called Cahotage in this book.
The most fundamental particles each consist of three spatial dimensions folded into a further real spatial-time dimension we have called Dq. In addition they have a touching point with Cahotage (which in normal particles is called enabler) and a superstring. The superstring consists of one normal spatial dimension and Dq (a line with a further Dq dimension) and is called activator.
The superstring has two important properties:
1) It is attracted to the enabler touching points with Cahotage and will allow the point to open up to Cahotage permitting spatial and time dimensions through to be used as work done (energy). This is an "end-on" reaction of the cross-section of the string (activator) with the touching point (enabler).
2) Because of the bending of space due to the contact with Dq the superstring repels other superstring sideways and also resists deformation. It was at one time considered that lines of force from a magnet could act in this way (Faraday).

Because there is space between superstring although it obeys the inverse square law approximately the correctness decreases the further away from the massive body. Thus space would be very flat not curved as theorised by Einstein.
The constant G could not be perfectly measured due to the problem of overlapping superstring from all directions and the limit on measurement.

At the edge of chaos new particles are collected or formed (continuous creation and creation fields... Hoyle).
In "normal space" the particles act upon each other by allowing motion in three ways
1) By the activator / enabler reaction ... this permits in slightly more space and time dimensions than appear to be required for our three dimensions thus causing time dilation (Einstein). Multiple dimensions of time may be used for complex acceleration, rate of change of acceleration etc. (imaginary time... Hawking).

2) By the bending of superstring storing the ability to do work or transmitting a wave (the activator / enabler action is required to set this off).

3) Apparent movement ... the bending of the path of "massless" particles such as the photon when they pass a massive body is due to the bending of space in Dq. The path of the light ray is not really bending .... it is taking the shortest route in multi-dimensional space (Einstein).

The smallest complete unit of a true particles would include one string and one touching point but they may have multiple strings and multiple touching points. Different properties may be visible at this level due to turning of the particle in Dq (Quark theory: the cause of up, down and strange is the turning of the faces into an "imaginary" spatial dimension !)

Different configurations of position of strings and touching points and their configuration in 3D.Dq space (multi-dimensional space) would explain the different "forces".

The inverse square law is maintained near to massive objects due to the superstring emanating in an even distribution around a spherical object. They stay apart because they repel each other sideways whilst deforming the three dimensions in Dq. This is folding of space or warping that is the cause of light bending round the sun (Einstein).The other laws of gravity are also all derivable but the quanta of gravity would be the reaction between one superstring and one Cahotage touching point (One activator/ enabler reaction).

The relationship between frequency of electromagnetic radiation (the number of strings that are close together) and Planck's constant (the Cahotage touching point) is also derived. The production of the waves is also due to an activator / enabler reaction and is also therefore subject to quanta. The dimensions of chaos let through would add to the disorder of our own universe as a form of Entropy that is called Extropy in this theory. This would be the cause of the arrow of time (A quantum theory of Gravity which includes an arrow of time: Penrose)

Massless particles such as the photon, Higg's boson and the gluon are not particles at all. They are all superstring ....either as waves or as single strings. They do not exist separately from a particle and seem to be a particle due to their activation of enabler points on any detector used.

There is a smallest limit (quanta) to the measurement of dimensions of space and time (already observed). This means that even if there is activity or there are objects beneath that level they cannot be correctly observed and will either fill a multidimensional quanta or they will not be observed at all. For example ... all electrons will appear identical because the small differences that might exist cannot be seen due to limitations of measurement.

Anti-particles (Dirac) will exist if made in this universe by violent activity. They will consist of an enabler on their superstring (i.e. opening point to Cahotage) and an activator on their particle ..... when they meet a normal particle they will open large rifts into Cahotage letting in large quantities of dimensions ("energy"). They will not be captured at the expanding edge of space with Cahotage because they will be destroyed along with the particle from this universe that attracted them, therefore there will be very few of them in the Universe.

Quantum theory involves uncertainty (Heisenberg). This can be both local and distant. These uncertainties are due to:
1) The activator / enabler reaction : The particle nature of photons is due to the enablers on the particles they excite. These particles are on the detectors and Chaos would not permit the reaction to be repeated exactly.
2) Chaos in the source of production. The exact activator / enabler reaction in the source of production can never be repeated
3) Quanta of measurement and inevitable manifoldation of the universe with a subsequent "partial volume effect".
4) The tight curling of space and the touching of different parts of three dimensional space due to Dq permitting waves and small particles to pass from one place in three dimensions to another with no apparent passage of time (Quantum tunnelling etc.) (Euan Squires mentions a ball of string in one line as a possible model!)

These causes of uncertainty will also apply to electron-microscopy, electronics and other areas of endeavour at the quantum level.

Since movement requires a balance of Cahotage (or chaos) and Order (the particle) it would seem likely that consciousness, which must require movement, would also require such a balance. Perhaps we can usher in an era of balance rather than an era of separated order and chaos.

Paul Goddard 1995

GLOSSARY

Activator: Gravity superstring, magnetic "lines of force", electro-magnetic radiation and the weak and strong interaction. See also force.

Dq: In this book Dq is the name given to the extra spatial dimension(s) required either as spatial-time or purely as spatial dimensions in order to permit the bending of the three normal spatial dimensions and in order that duality and complementarity is possible.

Duality: At the quantum level particles may have linked properties. When one property is known to a high level of accuracy the other property is only known to a variable probability level.

Enabler: The touching places on particles that permit them to contact Cahotage. They may be permanently open quantum sized holes through to Cahotage or they may be opened by the superstring

Extropy: This is similar to entropy and represents the build up of "used dimensions" in the universe. They may be incorporated as chaos or disorder in the system or they may be retained in Dq (this is not yet discernable).

Force. In Goddard's theory force is not used as a term. The nearest term to force is superstring or activator which activates the enabling touching spaces of particles to open to Cahotage and permits through dimensions of length and time (L^2T^{-2})

Grey Scale; Radio-density measurements of CT are displayed in a grey scale for a certain range of values. Above the range the pixel is white and alteration in value of radio-density make no difference to the grey scale. Below the range the pixel is shown as black and alterations in value below the range make no difference to the black pixel.

Manifoldation: the equivalent of three dimensional volume cell (voxel) and two dimensional pixels.

Partial Volume Effect : On computed tomography scans (CT) an object can be partially filling a voxel and the remainder of the voxel may be filled with another substance. The resulting combination results in a radio-density measurement that is shown as a grey scale on a pixel on a screen. The information within the voxel is then an average result rather than the true value of the object.

Pixel : A picture cell : as used on computers representing the smallest visible dot

Superstring ...this is the description of activator (previously known as lines of force).

Voxel : The three dimensional equivalent to a pixel